RIDER
BIOMECHANICS

Also published by Kenilworth Press

Ride with Your Mind Clinic
Ride with Your Mind Essentials

MARY WANLESS
RIDER BIOMECHANICS

KENILWORTH PRESS

Exercise can be dangerous, especially if performed without proper pre-exercise evaluation, competent instruction, and personal supervision from a qualified fitness professional. Always consult your physician or health care professional before performing any new exercises or exercise techniques, particularly if you are pregnant, nursing, elderly, or if you have any chronic or recurring medical problems.

The techniques, ideas and suggestions in this book are not intended as a substitute for medical advice. Any application of the techniques, ideas and suggestions in this book is at the reader's sole discretion and risk. The editors, author, and/or publisher make no warranty of any kind in regard to results from the information in this book. In addition, the editors, author and/or publishers of this book are not liable or responsible to any person or entity for any errors contained in this book, or any special, incidental, or consequential damage caused or alleged to be caused directly or indirectly by the information contained within. Kenilworth Press and Mary Wanless encourage the use of approved safety helmets in all equestrian sports and activities.

Copyright © 2017 Mary Wanless

This edition first published in the UK in 2017
by Kenilworth Press, an imprint of Quiller Publishing

British Library Cataloguing-in-Publication Data
A catalogue record for this book is available from the British Library

ISBN 978 1 910016 14 5

The right of Mary Wanless to be identified as the author of this work has been asserted in accordance with the Copyright, Design and Patent Act 1988

The information in this book is true and complete to the best of our knowledge. All recommendations are made without any guarantee on the part of the Publisher, who also disclaims any liability incurred in connection with the use of this data or specific details. All rights reserved. No part of this book may be reproduced or transmitted in any form or by any means, electronic or mechanical including photocopying, recording or by any information storage and retrieval system, without permission from the Publisher in writing.

Designed by Sharyn Troughton
Photographs by Peter Dove
Diagrams by Manoj Bhargav and Inger Recht

Printed in China.

Kenilworth Press

An imprint of Quiller Publishing Ltd
The Hill, Merrywalks, Stroud, GL5 4EP
Tel: 01453 847800
E-mail: info@quillerbooks.com
Website: www.quillerpublishing.com

CONTENTS

Foreword	9
Acknowledgements	11
Introduction	13

PART 1: THE FASCIAL NET AND FEEL — 17

CHAPTER 1 FASCIA — 18
An Introduction to Your Fascial Body	18
Slings, Sheets, Straps and Bags	21
Fascia in Horse and Rider	24
Tensegrity – Balancing 'Push' and 'Pull'	25
'Bounciness' and Hydration	28
Locked Long and Locked Short	29
Two Views of Training	31
Fascial Types	33
Myofascial Force Transmission	34

CHAPTER 2 FEEL — 38
Learning and Feel	38
Conscious Competence	40
What is Happening Inside the Body-Mind as we Develop Feel?	41
The Fascial Connection	42

PART 2: FRONT, BACK AND SIDES – THE OUTER SLEEVE — 45

CHAPTER 3 THE RIDER'S FRONT AND BACK LINES — 46
Stabilisation or Relaxation?	46
Front-Back Balance	47
The Superficial Back Line	48
The Superficial Front Line	50
The Superficial Back and Front Lines and 'Neutral Spine'	53
How to Balance the Superficial Back and Front Lines: Finding Neutral Spine	57
Breathing and Bearing Down – an Introduction to Your Core	62

How Different Riders Can Find Neutral Spine — 66
    Finding Neutral Spine as a Hollow-backed Rider — 66
    Finding Neutral Spine as a Round-backed Rider — 68
    Checking and Strengthening Neutral Spine — 71

The Superficial Back and Front Lines in the Thighs and Calves — 73
    To Elongate the Superficial Front Line and Shorten the Superficial Back Line — 74
    To Elongate the Superficial Back Line and Shorten the Superficial Front Line — 79
    Toes and Tendons — 81

CHAPTER 4 THE HORSE'S SUPERFICIAL FRONT AND BACK LINES AND RIDING IMPLICATIONS — 82

Hyperextension in Horse and Rider — 85
Rising Trot Mechanism – Ballistic Training for Horse and Rider — 87
    Matching the Forces in the Thrust of Rising Trot — 90

CHAPTER 5. THE LATERAL LINES IN RIDER AND HORSE – THE INTERMEDIATE AND OUTER STABILITY SYSTEMS — 92

The Lateral Lines in the Rider — 92
The Lateral Lines in the Horse — 94
    The Ribcage and Hind Legs — 97
    Bendy Body Parts — 98
Rider-Horse Interaction — 99
    Side Bends and Hot-air Balloons — 99
    Whose Problem Is It? — 101
    Know Your Enemy! — 102
    Narrowness in the Thighs and Seat Bones — 104
    Overcoming Your Instincts — 107
    The 'Boards' Exercise for the Intermediate Stability System — 107
    Using the Outer Stability System to Solve the 'C' Curve Problem — 112
    Influencing the Horse Who Falls Out on a Circle — 114
    When the Horse Falls In, or Wants to go Straight on — 115
The Lateral Lines in Your Legs — 116
    A Helpful Off-horse Exercise — 118
    An Overview: the 'Double Yellow Lines' — 119

PART 3: FUNCTIONAL LINES AND ARM LINES – PUSHING THE HANDS FORWARD — 121

CHAPTER 6. THE FUNCTIONAL LINES — 122

Working with the Functional Lines — 125
Off-horse Fixes — 127

Timing the Functional Lines	127
Centring the Cross of the 'X'	128
The Functional Lines in Your Legs	129
The Functional Lines in Your Horse	129
Some Thought Experiments with the Horse's Functional Lines	131

CHAPTER 7. THE ARM LINES — 134

The Human Arm Lines	134
Misbehaving Arms and Hands	138
Organising the Wrists and Elbows	140
Matching the Forces, Pushing the Hands	141
Passive Resistance	144
Short Reins and 'Arm Cuffs'	145
The Contact Scale	148
Shortening the Reins	149
When the Horse Puts a Loop in the Reins	150
When the Horse Offers to Lengthen His Neck	151
When the Horse Pulls	151
Visualising the Reins as Rods	154
The 'Arm Lines' in the Horse	154
Laterality	156
Grazing Position and its Effects on Asymmetry	156
Minimising Asymmetry	157

PART 4: TWISTS, TURNS AND THE REAL DEAL OF THE CORE — 159

CHAPTER 8 – THE SPIRAL LINES — 160

An Introduction to the Spiral Lines	160
Squeezing the Toothpaste Tube	162
Side Bends and Lateral Shifts	164
The Upper 'X' – Easing Your Head-carriage	165
The 'X' in the Abs	167
The Heavier Seat Bone	168
The 2.0 Problem	170
The Spiral Lines in the Legs	172
The Sacroiliac 'X'	173
The Spiral Lines in the Horse	174
And so We Finally Arrive …	176
Spiralling Around, or Not?	178
'Rebars'	179
Leg-yield and Shoulder-in	182

CHAPTER 9 – THE DEEP FRONT LINE — 184

An Introduction to the Rider's Deep Front Line — 184

The Deep Front Line from Toes to Diaphragm — 188
- Curling Toes and Wobbly Calves — 188
- The Thighs — 191
  - The balance between the inside and outside thighs — 191
  - The three orange segments of the thigh — 192
  - 'Leg-pits' — 193
- The Pelvic Floor — 195
- The Psoas Muscle – the Core in the Torso — 198
  - Functional and compromised psoas muscles — 199
- The Two Tracks of 'Locals' — 200
- The Quadratus Lumborum (QL) — 201
  - Intractable issues of mid-back and hips — 202
- The Diaphragm as Two Squares — 203

The Deep Front Line above the Diaphragm — 204
- Connectors and Spacers — 207

The Fan of Muscles — 208

The Deep Front Line in the Horse — 210

Core to Core? — 212
- Connecting to Your Horse's Psoas Muscles — 212
- The 'Ring of Muscles' and Longus Colli — 213
- The 'Treadmill' — 214

Conclusion – Riding as a Long-term Project — 216

Glossary – the 'Lines' — 218

Internet Connectivity — 219

Index — 220

FOREWORD
by Thomas Myers

I first fell in love with my wife watching her and her horse gallop in wild but coordinated abandon down a beach in Maine where I live. She and Dakota were one being; it was magic. Even though I was already developing Anatomy Trains at the time, I did not think of this event in terms of biomechanics – it was pure love – between her and the horse, and soon between her and me. Her connection felt entirely intuitive, and, non-rider that I was, I did not imagine that skill could be unpacked with analysis.

I first met and saw Mary Wanless at work coaching riding in an arena in the English countryside, near where I happened to be teaching a course. Here, many years later, I saw how wrong I was. Some of what I saw between my wife and Dakota that day was, of course, pure love, and parsing that is far beyond me. But here was Mary, clearly a master in her element, breaking the 'oneness-with-the-horse' part into its component parts, and using my 'map' to do it. It was a revelation.

Love remains a mystery for the poets and mystics, but the book you hold in your hand unravels the mystery of becoming one with your horse. Mary's many years of experience working with and watching every kind of rider from the beginner to the Olympian is distilled into this beautiful, practical, and comprehensive book for the rider who wants to inhabit that connection, that love, between human and animal.

Animal movement – and humans are animals – is not quite such a mystery as love, but the current scientific explanation – that individual muscles work via tendons over joints limited by bone shape and ligamentous restriction – is clearly inadequate to the task. The actual dynamics at play in riding – or any other complex movement – clearly require a new understanding beyond the levers and vectors of Newton's Laws of Force and Motion. I developed the Anatomy Trains myofascial meridians map to put the mystery of combined stability and fluidity of animal movement one step closer to a scientific analysis with a more relativistic understanding of how movement is generated and modulated.

Such an understanding requires some attention to the 'forgotten' body-wide fabric of connective tissue often termed 'fascia'. Seeing how the muscles function as coupled organs within the fascial net allows a new understanding of how agonists, antagonists, and synergists work together seamlessly to produce (or not, as when I myself began to learn to ride) the coordinated movement required to guide a horse.

Coupled with the work of the Danish veterinarians Vibeke Sødring Elbrønd and Rikke Schultz, referenced herein, who have done the yeoman work of extending my theories of myofascial connection to horse anatomy, Mary has intriguingly married the anatomy of the rider to the anatomy of the horse, and developed a method whereby many of the seeming 'faults' of the horse can be ascribed to the faulty biomechanics of the rider. This book details the corrections the rider can make to become one with the horse and correct the faults in both. It is a personal pleasure to see the Anatomy Trains scheme extended so beautifully, practically, and accurately beyond where I ever imagined it could go. It is a great professional pleasure to see Mary's many years of assiduous and detailed work take the form of this book, so that many others can appreciate the 'one-ness' that lies at the heart of the art, science, and craft of skilled riding.

Thomas Myers
author of *Anatomy Trains*
Clarks Cove, Maine

ACKNOWLEDGEMENTS

It was a chance encounter that led me to the work of Thomas Myers over ten years ago, and I instantly knew that his in-depth information about the workings of the body both validated my own intuitions, and went far beyond them. The scope and reach of Tom's work, and its influence in many different fields, is a remarkable tribute to him. I am one of many people with whom he has generously shared his time and expertise: I suspect he thought that this slightly mad English woman, with her request to collaborate on a series of articles about riding, would disappear into the ether and not follow through! But he stuck with me when the articles evolved into this book, and as a result, my work – and the field of rider biomechanics – is so much richer. Thank you, Tom, for your help with those initial articles, for reading the manuscript of this book, offering your corrections and insights, and writing the Foreword.

I am also grateful to Danish veterinarians Vibeke Sødring Elbrønd and Rikke Schultz, whose research is seminal to this book. Thank you Rikke for advising on the illustrations, and answering my many questions; I hope this book brings your work to a larger audience of riders than it might otherwise reach. I would also like to thank Professor Emerita Hilary Clayton and Dr Andrew McLean for their willingness to answer my questions. Both of them bring scientific thought into the traditions of riding, and also embrace the work of making that science more accessible to riders.

Thanks are also due to my illustrators Manoj Bhargov and Inger Recht for their patience through many revisions of the illustrations. Peter Dove has been both skilful and generous in our photographic endeavours for this book, with his daughter Millie Dove and her pony Tinker acting as wonderfully patient models (and for a fair proportion of that time, Tinker stood on four legs and Millie smiled!). I am also grateful to all of them for their participation in creating the videos that accompany this book (on www.riderbiomechanics.co.uk). I also want to thank Peter for being my 'go-to' person when I am stumped by my computer, and for being the technical wizard who masterminds my presence on the internet. This also includes mary-wanless.com, and dressagetraining.tv.

Many people play their part behind the scenes as I bury myself in writing, and in various other projects. When I am travelling to teach, I know that my home, horses, and riding centre are in safe hands. This is due to the wonderful Lise Twyman, accompanied by Karin Major and her team. Thank you.

A team of coaches working on five continents base their work on the contents of this book, and my previous books. I am so grateful that I am no longer a 'lone voice crying in the wilderness' and thank them for their faith in me, and for the hard work they have put into developing their skills. I know that they all have much more job satisfaction than they had when working from a more traditional knowledge base. Keep at it, all of you, and keep learning!

Thanks are also due to my publishers Kenilworth Press, and my editor Martin Diggle for their faith in my work and their patience with my timing!

INTRODUCTION

We live in exciting times. The last ten years or so have seen wonderful new discoveries in the realms of how the brain learns, and how our body-mind develops and applies the perceptions that underlay high-performance skills. (The term 'body-mind' makes clear that we cannot separate mind and body; they are one system, even though our everyday language implies that they are different entities.) We also know more about how coaching can enhance this natural learning process. Concurrently, we are discovering far more about the fabric of the body itself, busting biomechanical myths that have fostered misunderstandings in the riding world and beyond, in the wider world of exercise.

This new knowledge is creating a revolution in the understanding that guides skill development in yoga, martial arts, athletics, music, and other physical skills. With this knowledge we can be much more specific about the underpinnings of talent in riding, and how riders have to organise their own body in order to organise their horse. These leaves us all much less reliant on being born with talent, and more able to gain access to the (thus far) secret, instinctive knowledge of the talented few.

Many riders labour under the illusion that talent is a 'thing' that they somehow missed out on. The term acknowledges – indeed it legitimises – our inability to describe how the best riders utilise their body-mind. But talent is actually the *processes* of this, and new research suggests that it can be 'grown', through building skills step by step.[1] My experience of learning and coaching over nearly forty years supports this.

To replicate the processes of talent, we have to know what those processes are. The traditional horse world has suffered from a huge dislocation between the skills that comprise expertise, and how those skills are explained. Its language is fuzzy, open to misinterpretation, and often based on premises about human and equine movement that are now known to be misleading.

In order to learn to sit well and influence the horse like elite riders, average riders need access to a viable model of rider biomechanics *and* the rider-horse interaction. The latter explains how each partner influences the other. Through many years of empirical research (that is, research based on trial and error whilst working in the real world of riders and horses, and influenced as well by my scientific background) I have developed a model that has worked extremely well in practice. This has helped both average and world-class riders to improve their skills. There have been a fair few desperate riders in the mix too!

This client base has afforded me a unique overview of the challenges that riders face at successive levels as their skills evolve. Few challenges are unique to an individual – most are shared across the entire spectrum of riders. These challenges are overcome more or less easily according to each individual's tenacity and talent (for learning as well as riding). From this I have concluded that a viable model of riding skills *must be applicable to all riders*, and not only work for those who are either average or elite. The laws of physics and biomechanics are constant; they do not change between more or less accomplished riders, and more or less expensive horses. If we can only successfully address the issues faced by riders at one end of the spectrum, we do not have a complete model.

In my model of rider biomechanics, stability replaces relaxation as the rider's trump card. It is centred stability that gives elite riders the body control to remain still, elegant and effective on top of the moving horse. The skills of training horses then have a solid foundation. Without *inner stability*, we are all temped to create *outward* stability by pulling on the reins; but even this may not stop the bumps, wobbles, or jerks that the horse cannot possibly interpret.

Stable riders have reduced the 'noise-to-signal' ratio inherent in their sitting – they are neutral with respect to the forces of the horse's movement. This spares the horse from wondering which of the rider's movements are meaningful, and which are noise – a body part that has moved unintentionally, a loss of balance, or a desperate grab for stability.

Along with stability we also need 'feel'. Vision, hearing, taste, smell and touch on our skin are all more obvious to us than our sense of movement, which describes what happens under our skin – our kinaesthetic or proprioceptive sense from our muscles and sinews. All riders can develop feel; it is easier for some than others, but it does not require that we are born with talent. It simply requires learning how to pay attention in the most meaningful way. We could call the result of this learning 'kinaesthetic intelligence' or KQ. This joins your IQ and your EQ (emotional intelligence) to make a more rounded, accomplished you.

As a young professional, I spent hours thinking 'There's something my teachers aren't telling me', and reading every book I could find that might just have the answer. I now know that no one was deliberately holding back information, but that the skill of elite riders is so unconscious that it cannot be explained in words. This is why they can so rarely 'clone' themselves and pass on their skills to others.

It took many years for me to begin to discover the missing (unspoken) pieces in the puzzle, discovering how elite riders do what they do. We are going to begin our exploration into this by considering how our insides work, and how the fabric of the body can potentially allow us to generate both stability and feel. I promise not to blind you with science as I describe recent research which tells us that balanced riding starts not with our muscles and nerves. It starts with our biological fabric called 'fascia'.

If you think of your muscles like strings of sausages, then your fascia is like the sausage skin that links them into chains. Thus fascial fabric forms 'lines of pull' within the body that can extend from head to toe. Until recently, these have gone unrecognised by science; but they are now making their debut as they revolutionise our understanding of skill in many spheres.

Whilst struggling with your riding, you may have occasionally felt as if you were a puppet, with the horse as the puppeteer who was pulling your strings. If so, your flights of imagination were not so far from the truth, and whilst this represents the extreme of 'losing it', body control is hard-won for most riders.

Many of us find ourselves easily distorted by the horse's movement and one of my favourite questions is: 'Does the organised rider organise the disorganised horse, or does the disorganised horse disorganise the organised rider?'

Science has shown that we really do have 'strings' and this book is a map of the most important lines of pull in both rider and horse. Back through time, talented riders have always been able to balance the tension in their own 'strings', using this as the vehicle to reorganise their horse's 'strings', thus moulding his carriage and movement. But the secrets of that interaction were hidden in their unconscious minds – and also hidden in the secrets of fascia. Unlocking these makes the skills of ethical, influential, and elegant riding far more learnable and reproducible.

Regarding these lines of pull, within the chapter that focuses primarily on a certain line, its full name is given on first use. It is then abbreviated using the initials given in the glossary. When a line is cross-referred to in a subsequent chapter, its name is spelt out in full. My aim is to assist brevity whilst sparing you from feats of memory!

Welcome to a new dimension in your riding journey. It has the potential to enrich your skills and experience beyond measure!

PLEASE NOTE

This book follows the convention used in my previous books, and refers to the rider as 'she' and the horse as 'he'. This simplifies writing, and acknowledges that the vast majority of my readers are female. Within it, I mean no disrespect to male riders or female horses.

INTRODUCTION NOTE

1. See Coyle, Daniel, *The Talent Code*, Random House Books (2009)

PART 1 | THE FASCIAL NET AND FEEL

CHAPTER 1

FASCIA

An Introduction to Your Fascial Body

Fascia is often described as the 'cling-film' that wraps our muscles. You have probably seen it as the white film that surrounds the muscles of a chicken leg, or as the white lines dividing a joint of beef into segments (see Fig. 1.1), but fascia is so much more than this. It is a body-wide net that wraps not just muscle, but also every bodily organ, reaching from your skin to your bones. And it does not just wrap – this net pervades and invests nearly every nook and cranny with a kind of three-dimensional spider's web that keeps us and our cells together in recognisable form.

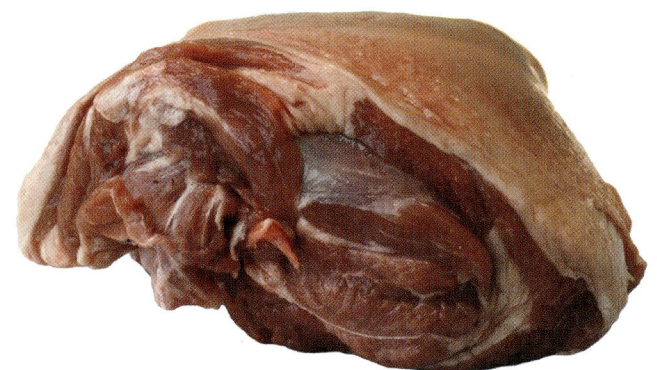

Fig. 1.1 (a) The fascial layer beneath the skin is loose (areolar), and provides a framework for fat cells - the net and the fat together make for a spongy feel. Beneath the skin is a thin, tough sausage casing for the whole body (the fascia profundis). More of this tough, glistening fascia encases each muscle, and runs both between and within muscles, visible in the cross-section as white fascial divisions. The smallest expression of fascia surrounds each muscle cell. If you could shampoo out all the muscle tissue, the fascia of this leg would hold its shape and resemble a finely grained loofah.

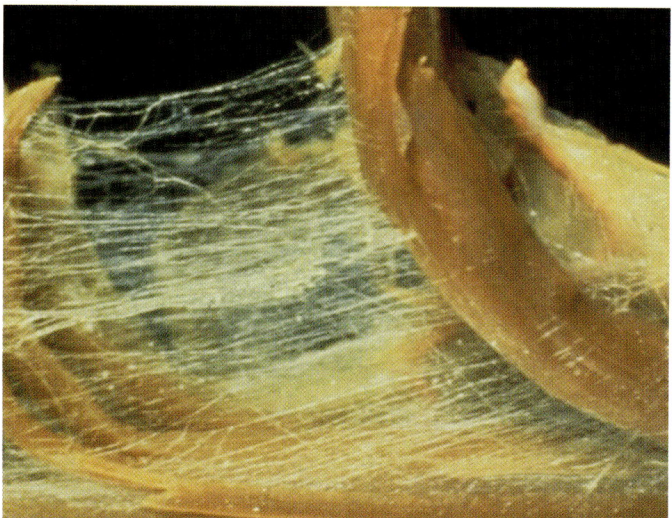

(b) A dissection of teased muscle fibres showing the surrounding and investing fascia. (Reproduced with kind permission of Ronald Thompson.)

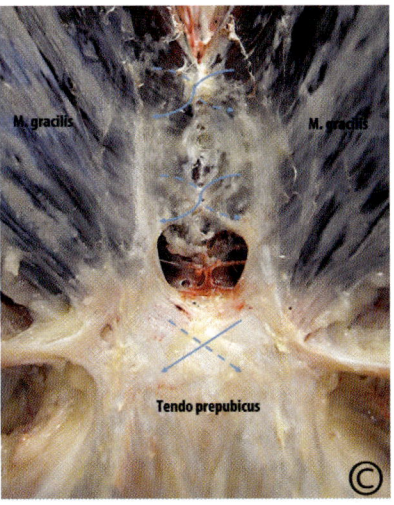

(c) A dissection of the fascia covering the horse's inside thighs, pubic bone and abdomen. Arrows highlight the 'Xs' in the fibres as they cross from right to left etc. This is part of the horse's Functional Line (see Chapter 6). (Copyright Elbrønd and Schultz.)

To get the basic idea think of an orange. It has an outer skin, and the inner skin of the pith (see Fig. 1.2). This connects to the central stem, just as your skin connects to your bones. Each segment of the orange is also wrapped in a skin, so adjacent segments are separated by two layers that can easily be pulled apart. Within the segments, each juice droplet is also wrapped in a skin. The orange consists of packages within packages within an overall package, in a system that is pressurised.

A juicy orange is a wonder to behold as well as to eat, and a dried-out one is such a disappointment. When this happens the pressure in each droplet is less; the segments become smaller – almost withered or wilted – and do not pull apart so easily. We humans are also more or less 'juicy', and the process of ageing has often been described as turning from plum to a prune. As the body dries out, its various layers and packages stick to each other, and the pressure inside our internal compartments becomes less. So whilst this book is primarily about riding, it is also about staying juicy, 'full' and youthful!

Fig. 1.2(a) When you cut an orange in half, you see its segments, but do not realise that two layers of 'skin' separate them. (b) The outer skin of the orange connects to its central core, as it does in us. (c) Separating the segments makes it clear that each one has its own skin.

Each of our muscles is wrapped just like each orange segment, and muscle cells are also wrapped, mirroring those droplets of orange juice. On the next layer down, a film wraps each individual muscle fibre, and at the end of the muscle these wrappings coalesce into a tendon. This in turn is continuous with the joint capsule and the ligaments that surround the joint. Thus it is the *same wrapping* that forms the next attachment for the next tendon, which leads to the next muscle, and so on. Thus muscles can be linked through the fascial fabric into long kinetic chains that can extend from head to toe (see Fig. 1.3).

However, the traditional techniques of dissection (and the anatomy texts that come out of them) are built on the premise that we have to divide the body into smaller and smaller parts, and look at the function of each individual muscle (discarding the fascia as we go). Thus the conventions of anatomy tell us that our muscles are separate entities, and that fascia is just 'packing material' – like the plastic 'peanuts' in your latest carton delivery.

Ever since French philosopher René Descartes thought of the body as a machine, we have tried to understand it by taking it apart. By studying each muscle, noting its origin, insertion, and direction, we isolated each one to its local function, and have largely lost the sense of how muscles work together.

Only recently has anyone gone beyond this 'isolated muscle theory'. It was a simple but radical move: by turning the scalpel on its side you can slide it between the fascial layers that would normally have been cut. Doing so reveals a body in which connected lines of muscle and fascia form long and specific lines of pull. As mentioned in the Introduction, you get the general idea if you think of these 'myofascial (meaning muscle plus fascia) meridians' as strings of sausages!

Thomas Myers, who pioneered such dissections, goes so far as to picture the entire body having 'one muscle that is divided into over 600 segments'![1] That's one complex orange – and just like that orange, we all need fascia to prevent us from becoming a puddle at our own feet!

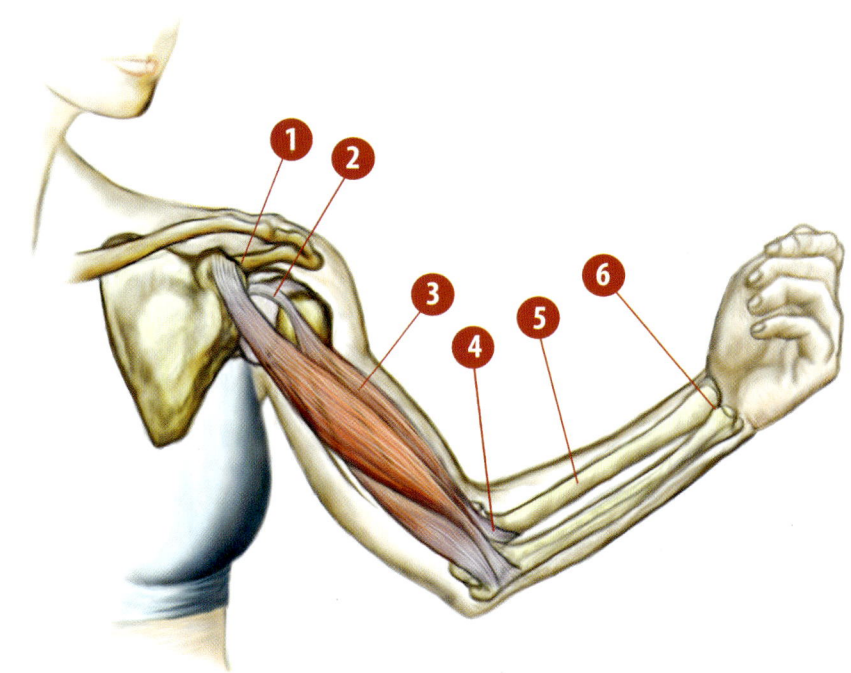

Fig. 1.3 The tensile part of mechanical forces is transmitted by the connective tissues, which are all connected to each other. The joint capsule (1) is continuous with the muscle attachment (2) is continuous with the fascia (3) is continuous with the tendon (4) is continuous with the periosteum that covers the bone (5), is continuous with the joint capsule (6).

Slings, Sheets, Straps and Bags

Remarkably, this resilient network of tough, stretchy gristle is stronger than steel, and even more remarkably, it is predominantly fluid. It consists of water and versatile forms of biological 'glue' or gel, in which tough strands of a fibrous protein called collagen are interspersed with more elastic fibres. This 'extra-cellular matrix' is 'manufactured' on the spot by fibroblasts (little fibre-making factories) which migrate through the spaces between the other cells and secrete collagen, rather like a snail leaving a trail. When more myofascial strapping is required in an area because of injury or repetitive mechanical strain, these cells get busy.

The texture of the resulting fascial fabric differs widely in form and function to provide the right balance between stability and flexibility. The weave is loose around individual muscle fibres (see Fig. 1.1(b) page 18), but more dense around muscles. Tendons and ligaments are more tightly woven, and the 'leather' that forms the basis of both cartilage and bone is densest of all. From the 'cotton candy' in your cheeks to the tough soles of your feet, fascia wraps your muscles and all of your organs from your brain on down, forming their inner structure too. It truly is a body-wide net, forming a unitary three-dimensional spider's web.

Movement – including riding – proceeds from an idea that comes out along the nerves to 'turn on' the muscles, so we talk of movement as 'neuromuscular'. But now we can see that those muscles are working within this body-wide fascial net, which *both directs and limits where the muscle's force can go*. So we might be more accurate to say that movement happens within the 'neuromyofascial web'.[2]

I am using the term 'fascia' in the singular here, because I want to emphasise the singular nature of this system. Only with a knife can we create the plural – fasciae. In medicine, the term 'fascia' is used to refer to certain specific dissectible bits of biological fabric – you have probably heard of the plantar fascia which forms the soles of your feet, or the thoracolumbar fascia in the lower back. As used here and in the parlance of therapy and training, however, the term fascia refers to the entire body-wide array of strings, sheets, sacks, bags, straps and nets that provide both environment and structure for the body's 100 trillion cells.

If you could dissolve everything else in your body, leaving only the fascia, you would still be recognisable as you. This three-dimensional web of fascia begins its life about two weeks into embryological development as a single structure – rather like frog spawn – that is folded and refolded in the complex origami of human development. Inevitably it will fray and dry out with age, and along the way it may be torn through injury, cut with a surgeon's scalpel or degraded by misuse, disuse, or abuse. But when you draw your last breath, it will still be one holistic system, expressing your own unique shape.

As Dr Robert Schleip points out, fascia has been the 'Cinderella' of body tissues[3] – systematically ignored, dissected out and thrown away, whilst being dismissed as passive and inert. It is not. Myers has identified twelve significant myofascial meridians, or lines of pull within the body. (These meridians were derived from Western anatomy, and are not the same as acupuncture meridians, even though both sets of meridians have some overlap.)

Through this process Myers is rewriting the anatomy books: we no longer have to rely on either the common but limiting 'isolated muscle theory' or its equally unhelpful new-age bromide, 'everything is connected to everything else'. Though this last statement is true – quantum mechanics shows us that everything really is connected to everything else – it is not very helpful in making clinical or training decisions that could improve your riding!

Myers' work has been seminal in bringing the concept of fascia into the public domain. His book *Anatomy Trains*[4] uses this term to name the 'myofascial meridians' map, and in his writing he makes the subject more lively and understandable by drawing an analogy to train lines. This includes 'stations' (commonly called muscle attachments), where the myofascial 'tracks' blend with joint capsules and therefore move bones.

Terms like 'local' and 'express' differentiate shorter muscles that cross only one joint from the longer, more superficial muscles, which cross more than one. Two or more locals will closely mirror the collective action of the single express, with the deeper locals having the greatest effect on our posture.

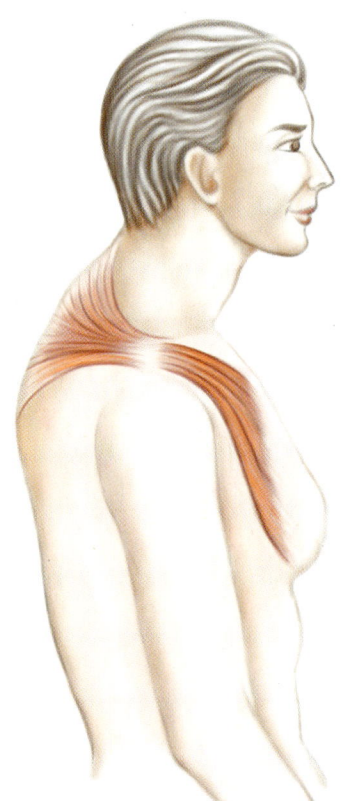

Fig. 1.4 A very common example of postural imbalance. When body segments are pulled out of place, muscles are required to maintain static positions where they are permanently contracted, or permanently strained.

Myers evolved his Anatomy Trains approach as a way of seeing, and teaching others to see, our postural imbalances and the resulting restrictions in our movement (see Fig. 1.4). As he says 'Stability, strain, tension, fixation, resilience, and postural compensation are all distributed via these lines.'[5] They become the actual tissue basis of the habits through which we use and misuse our bodies, creating visible patterns of asymmetry and compensation that we take into our riding. Here they become amplified by the forces of the horse's movement – not to mention the horse's habits too.

Myers has studied many forms of movement and manual therapy for over forty years, and his knowledge has evolved largely through learning, practising, teaching and developing his foremost modality, Structural Integration. This is a form of bodywork that specifically addresses fascia (rather than muscles or bones), and it was originated by Dr Ida Rolf (1896–1979).

Working hands-on to normalise the 'pulls' within our fascia can bring the entire system back into balance, reducing the need for repeated treatments that are really only addressing the symptoms of the problem. I love Myers' story of working on a client's back in Dr Rolf's class, using all the strength he could muster, and not achieving much. Then Dr Rolf – who was by that time a frail old lady in a wheelchair – came up behind the client and put her fingers each side of her spine. As Dr Rolf ran them through the tissue, the client excitedly exclaimed 'That's it, you've got it!' The client was unaware of the change in practitioner: Myers was left decidedly aware that the skill he was learning was not just a matter of strength, and that there had to be something else to it!

This notion might have a familiar ring to many riders, and Myers, like me, has spent years learning to understand and teach the subtleties of a skill that was not well-taught to his generation. As he developed and extended Dr Rolf's work he formed his own school, calling his work Kinesis Myofascial Integration (KMI). Primarily his school trains bodyworkers, but its reach has extended to teaching fitness trainers, Pilates teachers, yoga teachers, martial artists, sports coaches, and dance teachers about the myofascial meridians.[6] As Myers teaches these practitioners to observe postural imbalance, he uses simple exercises to show the myofascial meridians in action. He has had a worldwide influence, changing many a practitioner's view of the body they are training – a truly remarkable achievement.

Most of the interest in fascia was initiated outside of the academic and medical communities by the group that Dr Rolf trained. She was herself a researcher before developing her method, and she encouraged the scientific enquiry that could validate it. Some of these practitioners have furthered their cause by becoming academics, and an increasing number of scientists are now studying fascia as a remarkable and adaptable regulatory organ in its own right. They are researching its properties, and disproving the traditional view that it is nothing but an inert packing material.[7]

It turns out that the properties of fascia are indeed truly remarkable. The fascial system displays innate adaptability and 'intelligence', even without input from the brain or nervous system. As we shall see, however, it is a very potent source of information about the stresses it is experiencing, feeding this back to the brain via sensory nerves. Your brain cannot move the fascia except through the muscles; but the brain is listening very closely to what is going on in the fascial webbing. This is our richest sensory organ – there are more nerve endings in fascia than there are in the eye.

Remarkably, fascia transmits mechanical information through the body *at the speed of sound*, which is faster than electrical signals travel along nerves! Until recently, science and medicine have missed the way that, as mechanical forces change within the body, this information is communicated through the fabric of the fascial net, and particularly along the most significant lines of pull. We, as riders, subject our bodies to huge mechanical forces. The healthy functioning of our fascial net could not be more relevant!

Myers has focused his work primarily on mapping the most important lines of pull within the myofascial net, and whilst he has performed many dissections over many years, he does not consider himself a researcher. He believes that the practical work of manual and movement therapists runs ahead of the growing body of peer-reviewed research on fascia. Within the wider world, many are writing about – and have also developed classes in – fascial fitness,[8] fascial stretching[9] and fascial 'MELTing'.[10] They are building the momentum of approaches that are likely to become the next wave in the fitness industry – and also to extend beyond that, as we begin to understand the role that fascia can play in chronic pain, in how we develop and mature, and how well we age.

Fascia in Horse and Rider

Whilst equine and human anatomy differ, we are basically constructed in the same way, and of the same 'stuff'. In 2015, peer-reviewed research described the myofascial meridians in the horse. Danish veterinary surgeons Vibeke Sødring Elrønd and Rikke Schultz published their work[11] after performing the first dissections that searched for the lines. They wanted to elucidate the similarities and differences between the lines of pull in horses and humans, and it turns out that both are remarkably parallel.

The traditional understandings of the dressage world have intuited how some of the lines work; however, there are also many inaccuracies in our traditional ways of thinking. I am delighted that the chapters of this book, as they focus on one specific line, can draw on this research to talk about that line in the horse as well as the human.

Like all good bodyworkers, good coaches of rider biomechanics have a highly developed 'eye' for human bodies, and also equine bodies. They see how the distortions that pull a rider 'out of true' will also cause distortions in the horse and vice versa – for distortions in the horse's lines of pull will also distort the rider. My observation is that the difference between average and elite riders lies in the quality of connection and awareness between horse and rider – and this can be enhanced with conscious attention to each myofascial line.

If we apply what Myers calls a 'line analysis' to the posture, movement and function of horse and human, we discover how elite riders use their more resilient, well-aligned body to realign their horse, minimising the ways in which the horse can distort them. They balance the tension within their various lines of pull, training even tone through the lines to stabilise both themselves and their horse in a posture that allows controlled yet unconstrained movement. 'Equipoise' forms the baseline of good sitting in the rider, and good carriage in the horse. It occurs when all the lines of pull are balanced.

Most conventional trainers have a far better 'eye' for the horse's body than the rider's, and rarely treat both partners as a system. As with the other equestrian disciplines, dressage judging rates the horse's performance. The rider's skills are at least recognised and assessed in the collective marks awarded after the test. However, this still blinds the riding culture as a whole to power of the rider's biomechanics and her ability to influence the horse through influencing the interaction between (and the functioning of) her own and the horse's fascial nets. This book breaks new ground by treating the horse and rider *as one living unit.*

Whilst developing my skills as a rider and coach, I had begun to sense that muscles work in linked chains, and about ten years before starting this book I was thrilled to discover in *Anatomy Trains* a description of those very kinetic chains. I have benefited hugely from attending some of Myers' introductory classes, and from having bodywork with one of his practitioners. (I have even joined him in a dissection class, so see the human body for myself!) When Myers was able to watch my colleague Heather Blitz (an Olympic dressage rider with exceptional biomechanics) riding and teaching, he was able to offer his initial insights into how an elite rider's body operates, and how it influences the horse's carriage and movement.

Since then, our collaboration has bought rigour to my own work, grounding my discoveries about the rider-horse interaction within a theoretical framework, and defining the underlying structure of talent. It has also shown me that I was right to mistrust traditional explanations that use terms like tension, relaxation, suppleness, etc. whilst lacking a true understanding of their meaning and limitations. Myofascial meridians, and the properties of fascia, give us a much better explanation for talent and riding skills, including how those skills can best be developed.

Tensegrity – Balancing 'Push' and 'Pull'

Before Myers ever trained in manual therapy, he was an intern with Buckminster Fuller, the famous engineer and architect who brought systems theory to building design. His designs involved 'compression members' and 'tensional members' (see Fig. 1.5 page 27). The latter are strings that pull the compression members apart and hold them in resilient balance. Buckminster Fuller coined the term 'tensegrity' – a shortened form of 'tension integrity' to describe this type of engineering. So Myers entered the field of bodywork already thinking about tension, compression, systems, and how the body might or might not be designed in a way that mirrors traditional building design.

We often think of our skeleton as the part of us that resists the compression of gravity. If the body were indeed built like a traditional building, with one building block placed on top of another, the laws of physics predict that neither our lower back nor our knees could withstand the load pressing down on them. But within the body as a tensegrity structure, the tension of the fascial 'strings' can hold our bones slightly apart. Effectively our bones 'float' in the fascial net, without ever touching each other, except in injury or degenerative disease.

Tensegrity is very sensitive to balance. If the tensegrity system breaks down, so that its strings go slack in some places and tight in others, then the bodyweight compresses downward onto the joints, which come under huge pressure and become deformed over time. Less synovial fluid is produced within them, and the cartilage, which covers and protects the ends of the bones, can be worn away. A hip or knee replacement will be the ultimate answer to the excruciating pain that results from this. Our job, with the help of this book, is to nip these processes in the bud by improving the tonal balance across the body of both the rider and the horse.

If you catch the problem early enough, good training or (if necessary) bodywork, may well be able to restore the tensegrity of the system, so that the strings of the 'tensional members' pull the 'compression members' apart, thus taking pressure off vulnerable joints. This proves the value of distributing forces within a tensegrity system, and the vulnerability of a compression system in which forces become localised.

In the models of Fig. 1.5 the fascial strings are under even tension as they create the network in which bones are suspended. In the human body there is a further complication, since constriction or holding in one part of a line of pull can be passed along to other parts of that line. This can lead to injury at some distance from the source of the problem. Our own groin strains or a horse's flexor tendon problems may have come from imbalances further up or down that line of pull. Likewise, a rider's awkward shoulder position may come from how her pelvis relates to the horse.

Until a few years ago it was thought that tendons had no elasticity whatsoever; this is now known to be untrue. If we bounce up and down as in skipping with a rope, our calf muscles work isometrically, which means that they retain the same length whilst they contract. It is actually our Achilles tendons that change length and recoil elastically. This means that we, and our horses, are more like kangaroos than we had realised! We use the recoil energy that is stored in those tendons to propel us along, and elite athletes – especially sprinters – excel at employing the elastic recoil of the fascial tendons and ligaments efficiently.

However, fascia is elastic only up to a certain limit, which can be increased with the right training. Forced beyond that limit, it will deform plastically. Think of slowly stretching a common plastic carrier bag; the bag will not return to its original length, and if it is stretched too quickly it will tear. This is the source of most chronic injuries. Muscles, in contrast, are more biologically active: if you stretch them chronically, they will attempt to recoil to their resting length before, over the long term, they lay down more cells to increase that length.

Long-term habits of use, as well as scars from surgery or injury, or even psychological trauma, can cause fascial layers to bunch up and stick to each other like clinging plastic wrap. These fascial folds are pulled from somewhere else, following the specific lines of pull in the myofascial meridians. This causes restriction, misalignment, or mis-coordination up or down the lines, sometimes at quite a distance from the original problem.

Imagine two people sharing a bed: when one of them steals the bed clothes, who

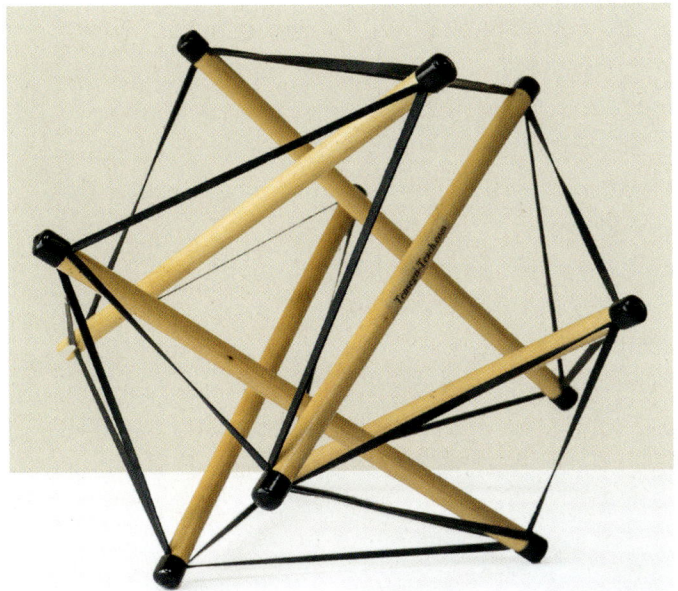

Fig. 1.5(a) A tensegrity structure comprised of rubber bands and wooden dowels. These are the compression members that 'float' without touching each other in a continuous 'sea' of balanced tension, created by the elastic bands.

FASCIA

27

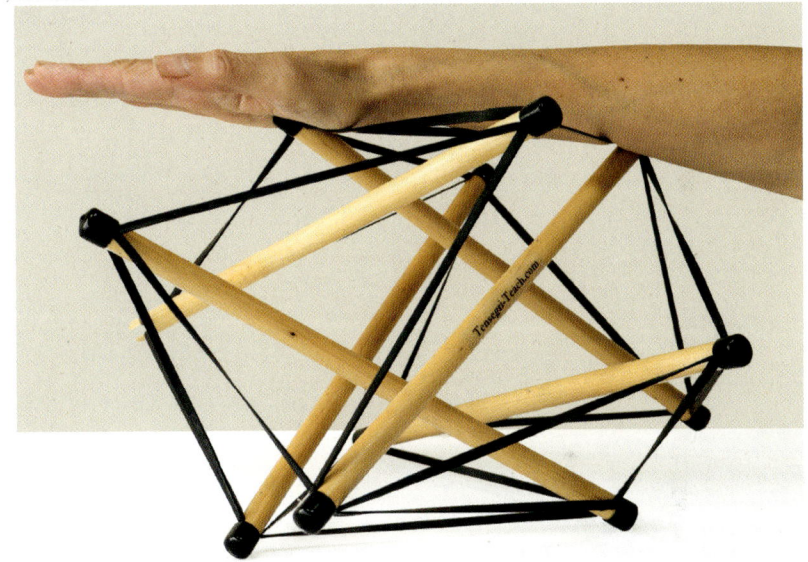

(b) A deforming force is distributed throughout the system, which bounces back when the force is removed.

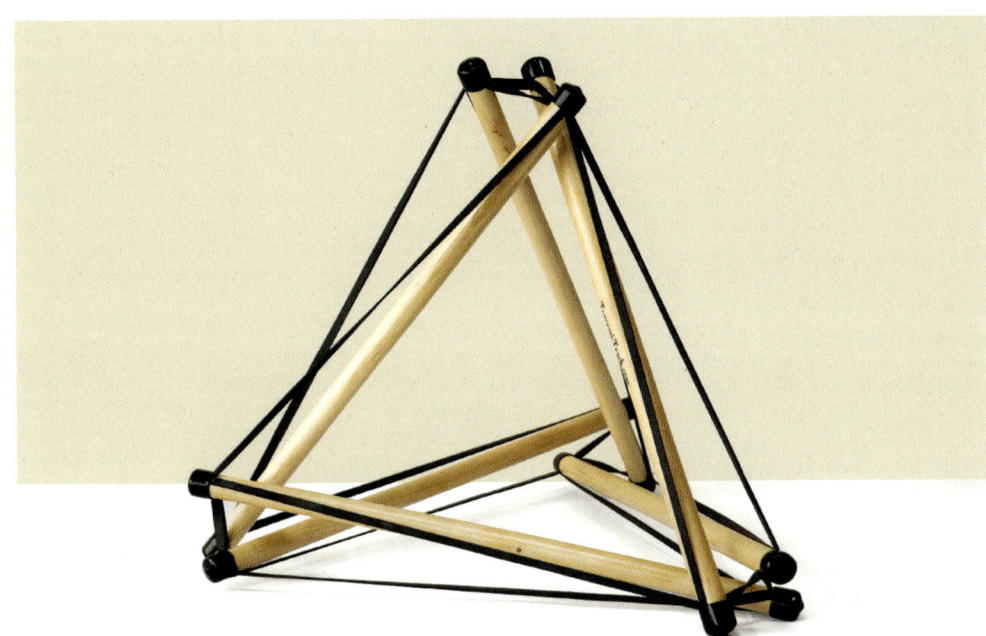

(c) Moving the dowels relative to the strings creates a form more like the human body. The long strings almost parallel to the dowels are like muscles; the shorter ones are like the ligaments around a joint. The latter are more likely to be damaged by a deforming force.

complains? It is the person from whom the bed clothes are stolen, who lies further along the line of pull – and this implies that the seat of pain can be far from an original injury. In practice, many clients want their bodyworker to work on the site of pain and solve it, but the point of view I am presenting here implies that the site of pain is the 'victim'. The real culprit (the 'criminal') can be elsewhere – and we do not want to punish the victims by mistake!

Myers tells the story of treating a wealthy man with a painful shoulder, who had already been to see every well-known bodyworker in London. The pain had not subsided, and a case history revealed that it began several weeks after surgery to remove his appendix. Knowing that the good and the great had already failed in their work on his shoulder, Myers took a deep breath and began to work on his appendix scar, releasing the tangled fascia. The man paid his fee with a grudging 'thanks for nothing'. Three days later he telephoned to apologise: his shoulder pain had gone.

A scar is like a snag within the body – be it human or equine. If you snag a knitted jersey, the pull follows a horizontal line. If you snag a hammock with a diagonal weave, the pull will go diagonally. The myofascial meridians map the most profound lines of pull, and you can think of them like the nap in a fabric, or the grain in a piece of wood. Thus within the *Anatomy Trains* model there are two important rules: a line of pull cannot make a sharp turn around a corner or move between different layers – the fascial fibres have to line up to create a viable pull.

'Bounciness' and Hydration

Healthy fascia is known for its white fibres, but it is in fact mostly water, like the rest of your body. Keeping fascia healthy is certainly dependent on nutritious food and 'nutritious' movement, but it is also a matter of keeping fluid moving through it. The ageing that lines our faces and makes our joints creaky is a breakdown in the fascia, caused by dehydration.

Young fascia has not only 'juice' but also 'crimp' – just like crimped hair – and this gives it the bounciness that we love to watch in young children, foals, and spring lambs. We lose this as we age, so it is literally true that when we fall off we 'don't bounce like we used to'. Fascia that is well hydrated retains its springiness better – think of the springiness in a natural sponge, which becomes stiff, or even brittle, when it dries out. Fascia is formed by genetics, nutrition, emotions (for example, stress or depression), environment, biomechanics and training, but the quality of fascia is influenced most of all by the way that it does or does not allow the passage of fluid.

The hydration of fascia changes its viscosity, which varies from thin and runny to more viscous and gel-like. It can potentially become extremely thick, or even dry and fibrous – just like the orange mentioned earlier. In fact, the ways that it can clog up and dry up in less healthy bodies (or less healthy parts of the body) are reminiscent of variations in the 'you-know-what' that comes out of your nose!

Fascia that is more watery allows various layers of the body to slide against each other. But as fascia becomes more viscous, different layers and structures stick more to each other, and the body (be it human or equine) becomes less responsive to training, since its layers are less able to slide and glide.

We all know that we should drink a lot of water, and that tea, coffee and sodas do not provide good hydration. (The advice is to take your weight in pounds, divide it by two, and drink that number of ounces of water per day.) However, that water might well go in through one end and out of the other without hydrating your fascia – if it is fibrous and dried out, water cannot get through it. This also means that metabolites (cell food) cannot easily pass from your bloodstream through the 'extra cellular matrix' of the fascia to reach the walls of cells that need nutrition.

Those cells then do not get food, and neither do they get water, as they would rather stay thirsty than 'drink' from the stagnant water that *is* available. The area becomes increasingly full of toxins, which are deposited within the already fibrous fascia, making it even more sticky. Think of this vicious cycle being like sand that is deposited in a river, with more sand being added as the river changes its flow around the growing sandbank.

Paradoxically, both inactivity and highly repetitive activity are dehydrating, since they both squeeze fluid out of the fascial 'sponge'. It is activity that promotes the flow of fluids, but *too much of the same activity* pushes the fluid out of the overused tissues. Rest after activity is what allows fluid to return to plump up and repair the dehydrated areas. When this does not happen, the result is likely to be loss of tensegrity and athleticism, creating poor posture along with chronic pain.

When different layers of the body are bound together, they have to be unbound again via movements like yoga or treatments like bodywork, which dissolve the 'fuzz'[12] and stickiness, restoring easy movement. Otherwise we carry these limitations around with us for the rest of our lives, affecting how our muscles work, what we can achieve, and even our psychological attitude. Understanding how this works turns some of our typical assumptions about muscles on their head, and also makes us realise that muscles themselves are not the real problem.

Locked Long and Locked Short

Muscles change from relaxed to contracted very quickly, and the fascia has to be able to accommodate these changes. Evolution has produced a fascial net that is built up from the very small – it wraps each individual muscle cell, and on the next level up, it wraps groups of muscle cells (called 'fascicles'). It then wraps each muscle, every organ, and the body as a whole.

At the smallest level, the netting that wraps individual muscle fibres does not follow the same line as the muscle fibres themselves. Instead, fascia has a double-lattice weave, rather like an onion bag, creating diamond shapes between the fibres (see Fig. 1.6 page 30). When a muscle contracts, the diamonds open to the side and come closer end to end. When a muscle is stretched, the diamonds extend, narrowing side to side, and the fascial fibres more closely approximate to the direction of the

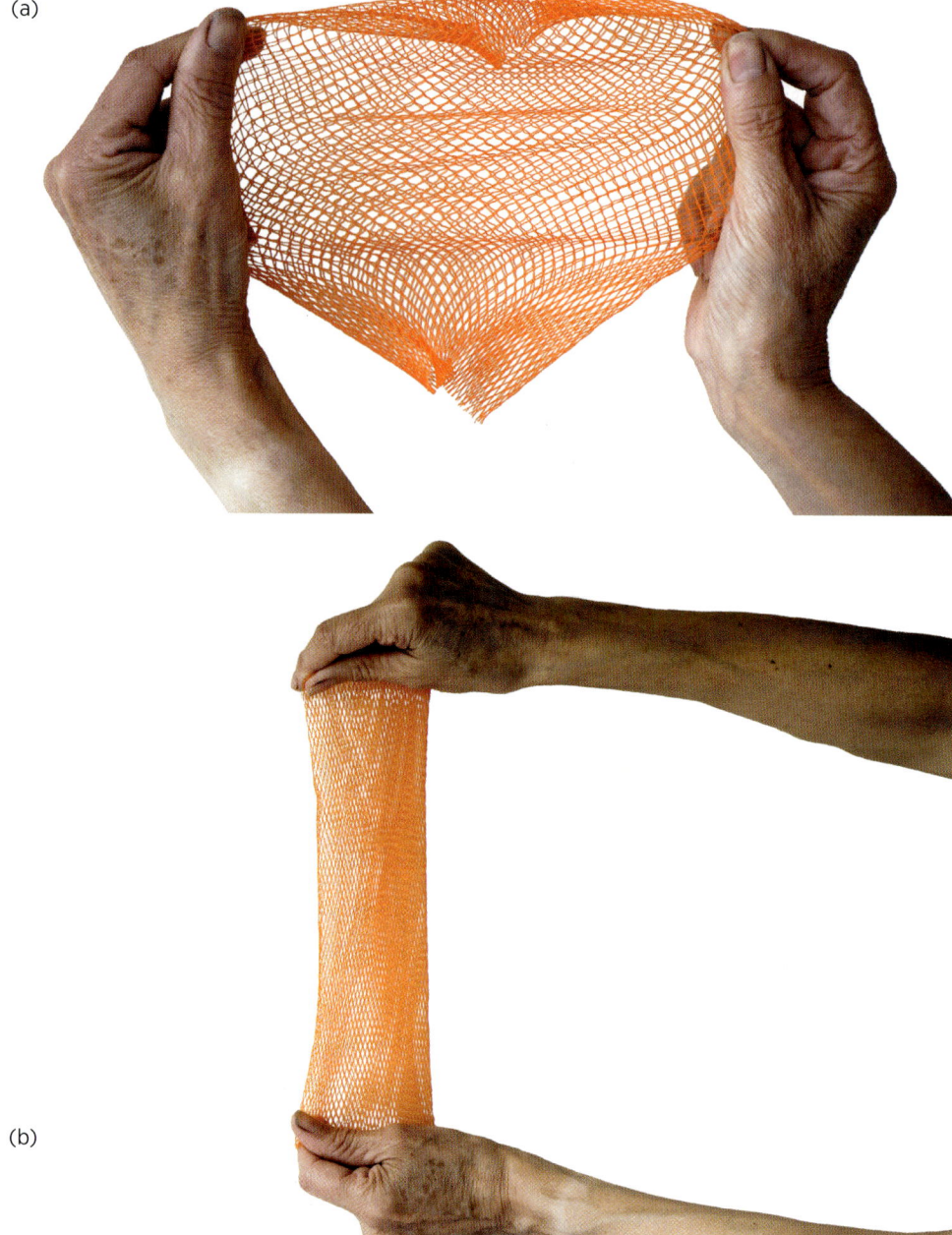

Fig. 1.6 The diamond weave of fascia makes it resemble an onion bag. Around muscles that are permanently contracted (a) it is 'locked short'; in muscles that are permanently strained (b) it is 'locked long'.

muscle. This design allows fascia to accommodate the normal contraction and stretch of all our muscles on a second-by-second basis.

As muscles contract and relax in succession they pump fluids around the body, and keep us stabilised for whatever activity we are doing. The problem comes – for both the fascia and the muscles – when muscles are held in a steady state for a long time, and stop pumping.

It is equally problematic whether the muscle is held in a shortened position, called 'concentric loading', or in a stretched-out but still tense state, known as 'eccentric loading'. In common parlance, both these muscles would be described as 'tense', but it is important to the rider, the therapist, or the trainer, to know which of these conditions the muscle is in. This determines the proper exercise, correction, or treatment. Interestingly, in this situation, both the eccentric and concentric muscles will show up as 'weak'.

Eccentrically loaded muscles – the overstretched ones – are often the ones that hurt. When your back, neck or shoulders ache and you go for a massage, the therapist often works with these sore muscles. But these muscles are actually working extremely hard to support the weight of your head and shoulders. Instead of contracting and relaxing in succession they are continually long and strained (see Figs. 1.4 page 22, 1.6). Because they are not pumping, these muscles become more toxic, dehydrated, and prone to annoying trigger points and chronic pain.

Whilst it is relieving to have the therapist massage the painful trigger points out of these muscles, it also makes these already overstretched muscles *longer still*. They might stop hurting for a while, but their extra length gives the muscles on the front of your chest the leeway to *shorten even more*! But these muscles are already in a concentric contraction, pulling your head and shoulders forward (as you slouch over your computer).

Inevitably, you will soon need another massage because your imbalanced tissues are locked into the spiral of eccentric loading in the back and concentric loading in the front. A longer-term fix for this problem is bodywork or training that will open the short muscles in the front and let the ones in the back return to their normal mid-range length. Balance will then be restored.

Myers uses 'locked long' and 'locked short' to label what happens in the fascia when we hold imbalances over time. A muscle in motion keeps the fascia around it supple, but a muscle held in one position will produce another kind of fixation in the diamonds around the muscle cells.

In the locked short muscle, the diamonds retract, and fascial bonding will tend to get stronger in the same direction as the muscle, where the points of the diamonds are closest (see Fig. 1.6(a)). In a locked long muscle, the fascial diamonds will get longer and narrower, and the binding will largely occur at right angles to the muscle fibres (see Fig. 1.6(b)). Simple exercise will not change either fascial condition, and real stretching – from yoga to Pilates to deep bodywork – is required to break up the limiting restrictions and allow free movement.

Two Views of Training

Many people use stretching and strength training in their attempt to address muscle imbalance, weakness, and pain; but because of their fascial restrictions they are only working on half of the problem. As in the shoulder girdle of Fig. 1.4 page 22, muscles are the victims of fascial architecture. People who stretch or who work out at the gym typically stretch or strengthen their body by targeting various muscle groups in turn, rather than taking a more holistic approach that focuses on lines of pull and stagnant areas. The solution may lie at some distance from the perceived problem, and is often on the other side of the body.

Training of individual muscles or muscle groups is obviously helpful, but not the whole solution. For one thing, any training of muscles has to be integrated into whole body function and aimed at an activity. Many athletes develop quadriceps strength through knee extension machines, where you sit and lift your leg against a resistance bar. This may strengthen the quads, but it weakens the body overall, since the quads are almost never used solo, but always in conjunction with the deep hip stabilisers on the opposite side.

Research is now showing that, throughout the body, holistic long-chain training works better in practice than muscle isolation training.[13] Exercises using Kettlebells are an example of the former, and are wonderful for riders, especially when you work with a coach who pays attention to your form.

Stretching has traditionally been viewed as a good practice, but now that assumption is more controversial. There are wide differences of opinion about its effectiveness, with many different 'recipes' being offered for duration, repetitions, the use of resistances, etc. But even without these controversies, the big problem is that few people stretch precisely enough, not realising that a small change in their starting position results in them *not stretching what they think they are stretching*. Since imbalances in your myofascial meridians are probably pulling you 'out of true', you will almost certainly *not* select the optimal way to begin a stretch. This is equally true when using exercise machines.

A fascial stretch feels slightly different from a muscle stretch and, instead of being localised, it is felt throughout the longer chain. Fascial stretches may look similar to stretches you already know, but they incorporate movement and are held for several deep breaths, during which you would be encouraged to focus on the feelings in your body. Traditionally, we stretch one muscle in isolation and in a static position; something we never otherwise do. A fascial stretch, however, follows the principles of normal movement, and maximises the benefits to fascia, muscles and nerves.

In addition to fitness in the muscular system, we must also consider 'fascial fitness'. Fascia also responds to training, although more slowly than muscle. You can build length, strength, and resilience into your fascial tissues (or not) in the way you train, and the way you ride. Fascial fitness primarily encourages improvements in rebound elasticity, so that we maintain the bounciness of youth. Exercises to regain elasticity in fascia are reminiscent of the ballistic stretching made famous by Jane Fonda. Skipping (jumping rope) and rebounding are good ways to start until you discover more options.

Preparatory counter-movement helps keep fascia healthy and organised. This, or 'the stretch-shortening cycle', are fancy terms for 'winding up' – we squat down before a high jump, and we bring our arm back behind us to throw or to swing a racquet. These movements 'pre-stress' the fascial net, as in drawing back the string in a bow before shooting an arrow. Think of how a horse bends the joints of his hind legs before he takes off for a jump.

Some of the taping that you see increasingly on Olympic athletes is being used to shorten a line of pull that might be vulnerable to damage. Self-treatment with foam rollers is also becoming common for rehydration and recovery. The MELT method (popularised by Sue Hitzmann) and other ball- or tool-based self-help programmes offer a treatment that you can do by yourself for yourself to address fascial dehydration and chronic pain.

There is a fitness revolution underway, an approach that 1) recognises and works with the fascial condition, and 2) develops body awareness and core muscle strength *across the myofascial lines.* Interestingly, our horses need exactly the same approach.

Fascial Types

Talented riders have their fascial net tensioned in just the right way; but how much of that is inborn and how much is learned? Researchers have only recently begun to study constitutional differences in fascia, and we can observe a spectrum of tightness or laxity in the fascial net of different people. We can make an amusing simplification by talking about 'Viking fascia', and 'Temple dancer fascia'. People with Viking fascia have a naturally stiffer fascial netting, whilst those with Temple dancer fascia have a naturally looser net. How you need to train depends on what kind of fascia you have.

Although there are formal tests, an easy way to tell where you are on this spectrum is to flex your wrist to 90 degrees, and then use your other hand to bend your thumb down to your forearm. If it comes close or touches, you are a Temple dancer. If it is still far away at full stretch, you are a Viking type. Neither is wrong or right, but it is good to know where you lie on this spectrum.

Vikings tend to love weight training and heavy work, whilst Temple dancers are often drawn to yoga and dance. In fact, both extremes actually need the opposite: Vikings will benefit from deep stretching, and Temple dancers will benefit from high muscle tone and oppositional balance in muscle groups. If you are Temple dancing rider, you might find joint stability hard to come by, as it far too easy to 'wiggle in the middle' as you attempt to sit the trot! A Viking rider will tend to be more stable, but might struggle to find the fluidity of easy sitting.

Age may be less kind to sedentary Vikings, but of course both types are subject to injury. When you cut your skin, the biochemistry of inflammation triggers the fibroblasts that manufacture collagen to transform into myofibroblasts, which can contract. This is nature's ingenious way of pulling the sides of the wound together to facilitate healing, and the resulting mechanical tension in the fascial net causes even more cells to become myofibroblasts. If the body overreacts and scar tissue is formed, this pulls on the fascial net even more strongly, in a process that causes yet more fibroblasts to become capable of contracting.

These changes also take place after injuries, like falls, that have not damaged the skin. It could be weeks after a traumatic event before you or your horse become noticeably stiff, with the body part most obviously affected being far from the site of the original injury. Unfortunately, myofibroblasts do not return to their original form – they remain contractile until they die and are replaced by new fibroblasts. Acupuncturists will tell you that they, as well as bodyworkers, have some of the best answers to this cycle, with their needles influencing the fascial net to reduce the pain and stiffness that this process creates.

Myofascial Force Transmission

Crucially for us and our horses, recent research has shown that fascia is the mechanism by which tensional force is transmitted through the body (compression goes through the bones and joints). This means that *fascia is fundamental to stability*. Good work on rider biomechanics trains fascia so that we become more stable as a whole, whilst also becoming more able to respond appropriately to small (and big) changes in our horses.

Whilst muscles respond to training in terms of days or weeks, fascia, even at its best and most healthy, responds in weeks or months – a process known as 'remodelling'. Myers calls the fascia 'the St Bernard dog of the body' – slow in responding, but reliable. It does not change in a hurry; but the more it changes, the more it can change. The tough part is to get the process started.

This next piece of information challenges our traditional assumptions about riding skills, and it allows riders to change in profound ways that would have been impossible within traditional thinking. Imagine each myofascial chain like a guitar string, which needs sufficient tension to play a note. It must be evenly toned all along its length to play properly, and the successful rider has cultivated that even tone. The opposite of integration is *isolation*, and areas of high-tone and low-tone tend to make for an unbalanced rider who lacks confidence.

The horse world is hooked on the idea of relaxation, and being told to 'relax' encourages people to think they should be loose and floppy. This makes their 'guitar strings' even looser. The 'loosely strung', low-pressure rider cannot control or sense what is happening to her body: any feeling at all will be fuzzy and unclear; in the worst-case scenario, it is as if this rider is a puppet and the horse is pulling the strings. Metaphorically speaking, her body cannot 'find the melody' (see Figs. 1.7, 1.8).

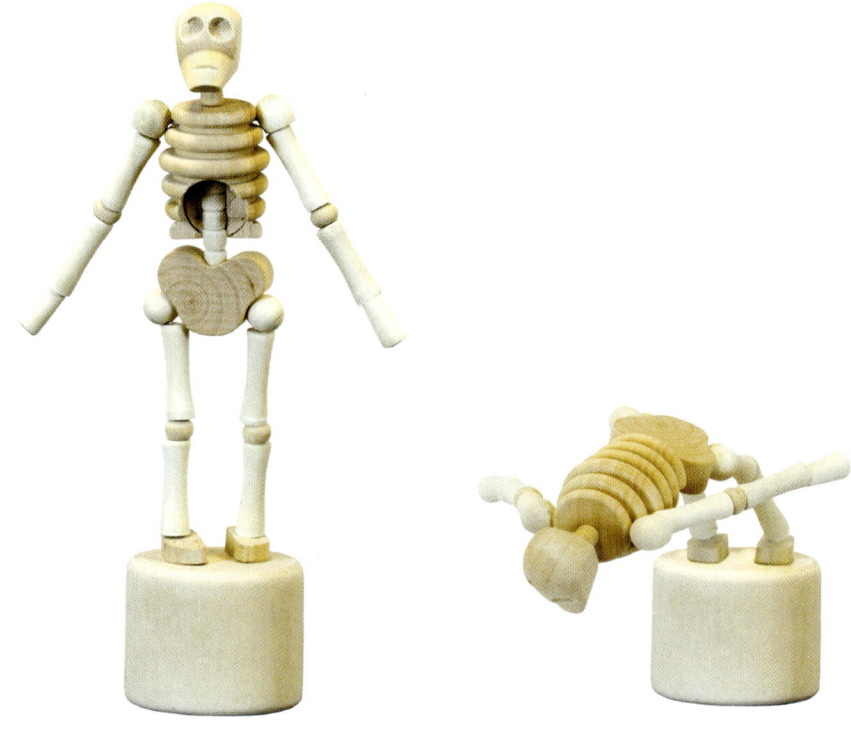

Fig. 1.7 This model shows how lack of tension in fascial lines of pull causes the structure to collapse, whereas the appropriate tension creates tensegrity and keeps us upright.

(a)

Fig. 1.8(a) Elite dressage rider Heather Blitz rides Otto, and both have the beautiful tensegrity of equipoise. (b) This rider is both leaning back and rounding her back, but the other key difference between her and Heather is how much floppier her body is. Both she and her horse are less well 'stuffed'. Follow her progress in Chapter 5 of *Ride With Your Mind Clinic*.[14]

(b)

The yoga expression 'tight is light' defines one of the benefits of an appropriately strung fascial net (think of a tennis racquet), since a floppy, 'noodly' body cannot support its weight well and is not resilient. Furthermore, loose tensegrity structures are easily deformed, so *they can change shape fluidly, but will collapse under significant load or impact.* As you tighten the strings (preferably evenly) you 'pre-stress' the structure, which becomes increasingly resilient, and able to bear more load and greater impacts without deforming. This more resilient body can sit to the trot better.

Riders are far too scared of tension, which is considered a terrible affliction. Riders look tense (I prefer the term 'brittle') when they are not breathing well. Of course a rider's joints should not lock up, but this is usually a protective mechanism that kicks in when the tissues are *too loose*. All human movement is a dance between the needs for both stability and mobility, and physical therapists increasingly understand that the body-mind has to *trust in its stability before it will allow mobility*. Thus it makes sense that legs and arms grab on when a floppy core cannot stabilise the rider.

The truth is that we cannot sit well without 'pre-stress' in the fascial net, and beyond deep breathing and a clear head, relaxation is not a concept that serves us well. Thus I often use the colloquial terms 'noodly', floppy, empty, or 'unstuffed' to describe how a rider's body looks and feels when there is a loosely strung fascial net, with various internal compartments at low pressure. Many riders can resonate with these descriptions – sometimes for their whole body, or for various body parts.

Please realise that it does not help us to think in judgemental terms like 'tense' and 'relaxed'. Once we view the body as a series of locked long or locked short tissues, we can work towards that easy, toned but responsive state. This requires a rider whose body is full, firm, toned, and 'stuffed' (like a brand new stuffed toy); only this enables a rider to sit so impressively still, which proves that the rider is *neutral relative to the horse's movement*.

When parts of our fascial net are 'noodly', we do not create the necessary internal pressure to generate stability. When we recognise a talented rider, we are seeing a body whose internal compartments are held at *strong internal pressure by a well-strung fascial net*. But we attribute such a rider's skills to relaxation, not pressure. Think too of the dressage horse in 'equipoise', or a stallion prancing in front of a mare; again you are seeing a body at high pressure, pushing out against the incoming tension of the fascial net. This is a world away from a 'soggy' horse whose rider has made his nose vertical!

When we are floppy, when we lose the balance between various fascial lines and antagonistic muscles, and/or when our fascial planes stick together, we are compromised as athletes. A high proportion of even advanced riders cannot control their lower legs in sitting trot, and when calves behave like noodles, the stabilisation of various myofascial lines is also compromised in the torso and thighs, right on up to the shoulders and neck. Thus coaching that develops body awareness and core muscle strength *throughout the lines* will improve rider functioning more than just focusing on any single body part such as the calves. In practice, the stability of the core and that of the extremities are intimately linked.

Finding a new 'feel' requires finding an alternative way of organising the lines of pull within your own riding body, and potentially within your horse's fascia as well. Each little discovery that one makes during a lifetime of riding refines how these lines of pull operate. This is a process that has no end; different horses challenge one's stability and organisation in different ways, and there is always scope for new discoveries. When we bring the human and equine bodies back into 'true', they become the beautiful, well-strung tensegrity systems they were designed to be. 'Equipoise' – in both horse and rider – is truly the art of dressage.

CHAPTER 1 NOTES

1. Myers, Thomas, *Anatomy Trains*, 3rd edn, Edinburgh, Churchill Livingstone (2014)
2. Schleip, Robert D. and Baker, Amanda, *Fascia in Sport and Movement*, Handspring Publishing (2015)
3. Schleip, Robert D. and Baker, Amanda, *Fascia in Sport and Movement*, Handspring Publishing (2015)
4. Myers, Thomas, *Anatomy Trains*, 3rd edn, Edinburgh, Churchill Livingstone (2014)
5. Myers, Thomas, *Anatomy Trains*, 3rd edn, Edinburgh, Churchill Livingstone (2014)
6. KMI practitioners and classes: www.anatomytrains.com
7. Fascial research www.fascialresearch.com
8. Schleip, Robert D. and Bayer, Johanna, *Fascial Fitness, How to be Vital, Elastic and Dynamic in Everyday Life and Sport*, Handspring Publishing (2017) www.fascialfitness.com
9. Frederick, Ann and Christopher, *Stretch to Win*, Human Kinetics (2006) www.stretchtowin.com
10. Hitzmann, Sue, *The MELT Method*, Harper Collins (2013), www.themeltmethod.com
11. Sødring Elbrønd, Vibeke and Schultz, Mark, Rikke, 'Myofascia – the unexplored tissue: myofascial kinetic lines in horses, a model for describing locomotion using comparative dissection studies derived from human lines', *Medical Research Archives* (2015) Issue 3
12. Hedley, Gill, 'The Fuzz Speech', www.youtube.com
13. Gracovestky, S., *The Spinal Engine*, NY, Springer Verlag (1989)
14. Wanless, Mary, *Ride With Your Mind Clinic*, Kenilworth Press (2008)

CHAPTER 2

FEEL

'Feel' a mysterious concept to those riders who suspect that they might be missing something, and the most wonderfully satisfying and exciting dimension of riding experience for those of us who are 'hooked' on developing it. However, for the talented few, feel can develop without them knowingly paying attention to their body. They can usually tell you what they changed in their horse, but not what they changed in themselves to achieve that.

Learning and Feel

This leaves a huge gap within our collective knowledge, fuelling the disconnection between expertise and explanation that is rampant in the horse world. The omission occurs because the learning of these elite riders has mirrored that of young children absorbed in play, who do not need to know *what* they are learning as they are learning it.

Childlike learning employs the nucleus basalis in the brain, which is switched on all the time in young children, enabling them to build their neurological 'maps' of the world effortlessly. This is the most efficient way of learning, and it has an evolutionary advantage in that children cannot possibly know what will be important in their future – so they are programmed to notice everything, in a semi-permanent state of fascination. Once this critical learning period ends, the nucleus basalis is only activated by something novel or unexpected, or by consciously paying attention – and this is the answer for adult riders.

Only a tiny percentage of children maintain this quality of focus, and they are likely to mature into extremely talented riders. The downside is that riders who did not know what they were learning as they were learning it will not be able to describe their experience in words! This reduces their ability to coach the average rider. The term 'expertise-induced amnesia' has been coined within the academic literature on expertise, and it suggests that experts inevitably forget things they once consciously knew.[1]

However, the notion of 'amnesia' barely hints at your experience if you sat on your first pony with the advantage of a well-strung fascial net, good breathing, and natural confidence. Both then and later in life, the 'A, B, Cs' of riding will elude your conscious mind since you did them so naturally – you are rather like the proverbial goldfish who would never discover water! When you teach in years to come you might start, for instance, from 'H'. With the addition of some real amnesia for skills you developed as time went by, you might even begin from 'X'.

One prominent researcher has gone so far as to call these omissions 'the curse of expertise' – even though both elite and average riders rarely imagine that expertise might bring with it any downsides, let alone a curse![2]

As I see it, the 'curse of expertise' has had a huge influence on our entire riding culture. It created – and it continues to preserve – the dislocation between expertise and explanation. Sayings like 'grow tall and stretch your leg down', and 'stick your chest out', were first said by expert riders, representing in language discoveries that can only be made though time and talent. This means that they, and other traditional sayings, are actually 'Xs' presented as 'As'.

When beginner riders hear these statements, they make the obvious interpretations, creating a host of problems. The true meaning of, 'Grow tall and stretch your legs down', for instance, involves the Lateral Lines of Chapter 5, not the much more intuitively obvious front of the body. But few coaches realise that they are perpetuating a massive misunderstanding, or see the problem they have created. It is no wonder that so many riders get stuck and confused.

It took me many years to travel a path from 'A' to 'X' and discover what these traditional sayings actually mean.[3] For me, teaching the real 'A, B,Cs' of feel and skill begins when I first put my hands on a rider's body, usually to realign it, or to offer a resistance (see Figs. 3.6 page 58, 3.7 page 61). The question 'What is this like compared to normal?' invokes a comparison – and all feelings are comparisons to a previous norm.

These comparisons are crucial. The many riders who are stuck on a plateau are missing them – they are always doing what they always did, and always getting what they always got. In the same vein, when you go into a room that smells, you will notice the smell strongly at first and then gradually habituate to it. Only by going outside and breathing some fresh air do you give your nose a contrast that will make the smell evident when you re-enter the room. This contrast allows 'news of difference' to enter your system.

After the initial novelty wears off, most riders lose any 'news of difference', and land on a plateau that is composed of some rather unsophisticated habits. K. Anders Ericsson, one of the foremost researchers into expertise, has described this using the wonderful phrase 'premature automation'.[4] A pattern that is far from optimal has become automatic – it is a habit that prevents the rider from improvising, and searching for a better way.

Ericsson considers this as a state of 'arrested development', and far beneath the skill level that is available to everyone. Sadly, the plateau is the fate of so many riders that we virtually consider it normal! Even more sadly, these bad habits are 'written in' to the nervous system, eventually causing remodelling responses in the fascial system. Thus they become embedded ever more deeply – in the tissue as well as the brain.

Skilled riders express the same issue in a different way. Ask what one is doing, and the answer may well be 'Nothing'. What this really means is 'Nothing that I register as unusual', which is equivalent to saying 'I don't smell it anymore'. The difference between an elite rider, and a rider with 'premature automation', is that the elite rider has ingrained much more useful habits, and can improvise by drawing from a varied 'toolkit'. But you should realise that, if you hear the response above and attempt to do 'nothing', your 'nothing' (which will probably be an attempt at relaxation) could be a million miles away from the elite rider's 'nothing' – which is, in fact, a well-strung fascial net that is no longer registered!

As you are sitting reading this, notice your tongue in your mouth, and your feet on the floor. Notice the movement of your ribs as you breathe. This is feel – your perception of your body in motion, known formally as kinaesthesia. With practice, your sense of feel in riding can encompass how your backside and thigh touch your horse, and how he touches you. You become able to influence the surface he gives you to sit on. You become aware of his body out to its extremities. As a driver, you can (I hope) extend your 'feel' in the same way to include your car, enabling you to park it in a rather tight space.

Conscious Competence

Not enough riders realise that every riding breakthrough begins when you notice what is actually happening *now*. This is the famous 'Aha!' moment, which takes you from being 'unconscious of your incompetence' (the state where you do not know what you do not know) to being 'conscious of your incompetence' (where you get to know what you didn't know). This can be shocking; but the realisation that 'I just curled my toes!', 'I stopped breathing' or 'I hollowed my back' is enormously empowering. As your skills improve, you begin to notice how your horse played a role in disorganising you, and you learn how to beat him to it, making a correction before you both go wrong. After enough repetitions new pattern begins to function in 'unconscious competence'.

Riders develop skill and feel most effectively when they work at the 'edge' of their existing skill set, and practise 'getting it' and 'losing it'. This is the state of being 'consciously competent'. If you had arrived at 'unconscious competence' *without going through* 'conscious competence' (like a child at play) you would be a supremely talented rider who wonders what is wrong with the rest of us! We have to proceed laboriously, from unconscious incompetence to conscious incompetence, and through conscious competence to finally arrive at unconscious competence. But take heart from the elite golfers and tennis players who talk publicly about 'rewiring' their shots through this process! The best coaching invites even elite riders to do the same.

K. Anders Ericsson used the term 'deliberate practice' for the stage of 'conscious competence'. It was termed 'deep practice' by Daniel Coyle, and popularised in his book *The Talent Code*, which I highly recommend.[5] Ericsson's research suggested that it takes 10,000 hours of deep practice to become elite at any skill. That means noticing for four hours a day, five days a week, fifty weeks of each year for ten years! Few riders come close to this, in either time, or focus.

It is important to realise that mindlessly riding round and round does not count, and also that practice makes perfect what you are *actually* practising, not what you *ought* to be practising or *wish* you were practising. Sadly, you can practise bumping until you bump permanently (and perfectly)! In addition, you cannot improve by only practising what you already do well.

'Checking in' occasionally does not count either and, unfortunately, the adult brain is easily tripped up by thinking about the past or the future, perhaps with special emphasis on what *might* happen. You might be wondering who is watching you and what they are thinking, or whether your horse will spook at something. You might be watching the scenery. You might be planning tonight's supper. Or you might be trying harder, with all the angst and willpower that this implies. Even distractions that appear to be external exist primarily inside our heads (only we gave them that status), and perfect (deep) practice is a meditative art form. Simply noticing your body's perceptions is a primary skill for riders.

If a riding career spans thirty years of deep practice, this means that the only thing that changes over time is the succession of 'its' that you get and lose, and the sophistication of the nervous system through which you process those 'its'. There is nothing else! This explains how, after about fifteen years of us working together, US International Grand Prix dressage rider Heather Blitz could say to me, 'It always amazes me how the changes we are making now, which are so minute compared to those early lessons, still feel to me just as huge and meaningful.'

She is processing them through the kinaesthetic version of a stronger microscope lens, perceiving differences that she could never have dreamed of all those years ago. Her feel and her 'kinaesthetic intelligence' have developed enormously over time, and that acts to magnify her perceptions. For her, every ride is deep practice, and she has rediscovered the fascination of the young child who is learning whilst playing.

I like to think of the 'movement brain' like a manual camera lens. In under-focus, a lot of information is perceived as a blur. Just the right focus brings clarity, but over-focus means that we 'cannot see the wood for the trees'. This is what happens in *trying*. For my generation, who grew up on 'if at first you don't succeed try, try again', this has been a scourge. However, young people today are equally likely to be under-focused. Without the focus that keeps you noticing in every stride, there is no meaningful (deep) practice.

What is Happening Inside the Body-Mind as we Develop Feel?

Even after a long time spent learning in this way, and even for an elite rider, new breakthroughs can be really surprising. They can 'come from left field', with a logic that makes sense in hindsight, especially when viewed through the 'Anatomy Trains' framework. But it was not predictable. After all, if you had known where to look for the answer, you would surely have found it by now!

What *is* predictable is that one's sense of feel begins with the ability to perceive and change a small body part. But as one's feel evolves, each small change will include a more holistic, body-wide sense of how that change is actually fixing a weak link within a longer myofascial chain, and/or creating balance and stability between different chains. Since any chain is only as strong as its weakest link, a tiny adjustment to that disconnected link can make a huge difference, first to the rider, and then to the horse.

The Fascial Connection

So what is actually happening in the body-mind, especially as these longer chains become more 'feelable' and functional? Traditionally, textbooks on proprioception would teach you about muscle spindles, which send messages to the brain about changes in muscle length. They would also tell you about the Golgi tendon organs, which send feedback to the brain about changes in muscle tension. But recent research has shown that these are not the big players. To understand these, we need to know more about fascia.

Recent estimates have shown that for each muscle spindle, there are about six receptors in the surrounding fascia. This means that your brain is receiving *six times as much* feedback from your fascia as it is from your muscles! As we have said, fascia is the richest sensory organ in the body. To the Golgi tendon organs, we can add the Pacini, Ruffini and interstitial nerve endings. The Pacini endings measure pressure, and the Ruffini endings measure the shear forces between different layers of fascia. The interstitial endings measure all of these forces (though less precisely), and also register pain.

As Van der Wal showed in his research, 'Nerve endings arrange themselves according to the forces that commonly apply in that location in that individual, not according to a genetic plan, and definitely not according to the anatomical division that we call a muscle.'[6] Through these various nerve endings the brain is registering how and where fascial connections are being made – how those lines of pull are operating. Through feel we register pressure changes and shear forces in our own bodies, and potentially in our horses'.

Athletes have more of these nerve endings than the general population, but researchers do not yet know whether having more nerve endings makes them better athletes, or whether athletic training increases the number of nerve endings. Having seen so many average riders develop their feel, my bet is on the latter. Either way, to quote fascial researcher Robert Schleip: 'The nervous system is intently listening to the fascia.'[7]

The myofascial meridians link muscles into functional stabilising systems, and research has now proved force transmission from one muscle to another via the connecting fascia. But fascia transmits force *only when taut*. A floppy 'noodly' rider has her internal pressure low and her 'guitar strings' too loose; so instead of transmitting and redirecting the forces of the horse's movement, this rider is 'bobbled about' by those forces. This makes for a fuzzy rather than clear sense of body – too much noise, too little signal.

In contrast, the talented well-strung rider has the increased feel and skill that come with increasing perceptual refinement of the shear, gliding, and tensioning motions in the fascial membranes. Remember, however, that 'expertise-induced amnesia' will almost certainly leave such a rider unable to tell you what took place in her own body to attain a certain change in the horse, even though she will be aware of that change in him. But with our new information about the fabric of the body, and with an appreciation of the omissions caused by 'expertise-induced amnesia', we can delineate so much more about the 'how' of riding. Aspiring riders can thus develop both skill and feel.

The challenge of learning is to become aware of, and to exert influence over, the distortions within this series of myofascial lines. We train our horses by training ourselves, for controlling and balancing our own lines of pull is the key to influencing the horse so that we balance his lines as well. It is these concerted muscle chain connections in horse and rider that that determine whether we organise him or he disorganises us.

For those who coach, the knowledge of the Anatomy Trains enhances the ability to diagnose the source of problems, and brings skill at finding appropriate 'next-step fixes' for riders at all levels. A 'line analysis' of musculoskeletal functioning in both horse and rider is a very effective way of progressively elucidating the 'how' of riding, and of developing the skills to train malleable and responsive horses. The emerging field of rider biomechanics now has a firm base in an established theory.

CHAPTER 2 NOTES

1. Beilock, S. L. and Carr, T. H., 'On the fragility of skilled performance: what governs choking under pressure?', *Journal of Experimental Psychology: General* (2001),130, pp.701–25

2. Hinds, P. J., 'The curse of expertise: the effects of expertise and debasing methods on the predictions of novice performance', *Journal of Experimental Psychology: Applied*, 5 (1999), pp.205–21

3. Wanless, Mary, *Ride With Your Mind Essentials*, Kenilworth Press (2001)

   Wanless, Mary, *Ride With Your Mind Clinic*, Kenilworth Press (2008)

   Wanless, Mary, *The Naked Truth of Riding*, DVD set, Self-published (2014)

4. Starkes, Janet L, and Ericsson, K.A. (eds), *Expert Performance in Sports: Advances in Research on Sport Expertise*, Human Kinetics (2003)

5. Coyle, Daniel, *The Talent Code*, Random House Books (2009)

6. Van der Wal, J., 'The architecture of the connective tissue in the musculoskeletal system: An often overlooked functional parameter as to proprioception in the locomotor apparatus', in Huijing P.A. et al. (eds), *Fascia Research II: Basic Science and Implications for Conventional and Complementary Health Care*, Munich, Germany: Elsevier GmbH (2009)

7. Schleip, Robert and Baker, Amanda, *Fascia in Sport and Movement*, Handspring Publishing (2015)

PART 2 | FRONT BACK AND SIDES – THE OUTER SLEEVE

CHAPTER 3

THE RIDER'S FRONT AND BACK LINES

Stabilisation or Relaxation?

So many riding teachers and trainers will tell you: 'Relax!' Whilst there is some truth in this, there is also a big lie.

An apparently relaxed and responsive rider demonstrates the trickiest optical illusion of skilled riding. Whilst appearing relaxed, she is actually in a state of incredible mental focus and, physically, is drawing on a deep well of strength that few riders even dream of. However, with the help of the programme explained in this book, most riders can, in fact, find this core strength and adaptability. The key to this wellspring is tone and balance, not just in muscles, but between the various myofascial lines. Balancing opposing lines and muscle groups with each other gives each rider greater stability and range of movement. Perhaps we could coin the term 'myofascialature', since this includes both the tone of muscle tissue and also the tone of the sinews which join muscles like links of sausages. Both are essential to apparently 'relaxed' (beautifully stabilised) riding.

In other words, it's not enough just to follow our riding tradition by thinking 'relax', nor (if you subscribe to the idea that more muscle power is needed) to go to the gym and tighten those wobbly abs. Even tone along the lines is a more important – but largely unrecognised – attribute than strength in any given muscle. Just for starters, the organised rider pits her back against her front and her right side against her left side. What is now commonly known as 'core strength' – addressing the essential muscles in the middle, like your pelvic floor – ironically depends on balanced tone in the outer 'sleeve' of musculature.

Therefore, we will explore bringing tensegrity balance to these outer lines first, because without that balance in place, even wonderful tone in the core muscles cannot easily express itself as responsive movement. (Too many people in the personal training community have spent so much time and effort strengthening their 'core' that they have become rigid, manifesting poor movement the second they step out of the gym.)

Think instead of an acrobat, seemingly relaxed but without an ounce of floppiness! This quality is obvious too in skilled equestrian vaulters; but when a dressage rider places herself and her horse in 'equipoise', few people recognise the positive and encompassing tension they share. To contrast this with relaxation, remember the collapsing skeleton toy of Fig. 1.7 page 34. When you push up from beneath this, all of the strings go slack – the result is relaxation, but no tensegrity, and this is not our ideal!

Adaptive stability, not floppiness, is the bodily expression of confidence –we could call it physiological confidence. Many, many riders, when they have learned to balance the length and tension within the front and back of the body (whilst also breathing – stay tuned, we will get there) have said to me 'I feel more stable, still, connected to the horse, and so much safer.' It is sad that so few coaches realise that, for many riders, these physical skills need to come *first*, generating an effective riding toolkit which, in turn, engenders confidence. (We are not speaking here of the small percentage of gung-ho dare-devil riders who think they are invincible, whose confidence outweighs their competence!)

There is a gift that comes from basing the theory and practice of rider biomechanics in the concept of tensegrity and the connecting myofascial meridians. It is the realisation that the greatest benefits will often come from bringing the system into balance by increasing its 'pre-stress' (myofascial tone). This happens when we *shorten the overlong parts of the myofascial lines, and lengthen the overly short parts.* As you progress through this book you will learn a system of isometric resistances (either using a hand on the saddle, or as offered by a coach or friend) that induce tone in the longer, more 'noodly', lines, as well as some stretches for the chronically short parts. You will soon discover how phenomenally effective this is.

We will start with the front and back lines that originate literally on the top and bottom of your toes and reach up to hug the chest and spine. Then we will move to the sides, where the two halves of the ribcage are connected to the outer legs, forming our first 'stability system'. Then we will move into the core – our deeper stability system – and discover how to connect it all up.

Front-Back Balance

This all sounds good, but are we talking about something real, or something we imagine? The answer is that we are talking about something real, but only just being recognised. The connections made by the Anatomy Trains are very real, as evidenced by the dissections of Myers and the Danish veterinarians.

We begin our look at how the Anatomy Trains work in riding by focusing on the Superficial Back Line (SBL) the Superficial Front Line (SFL). Both of these are essentially continuous lines of muscle and biological fabric (myofascia) from the top and bottom of the toes respectively all the way up the body to the head.

Whether they frame it this way or not, almost all riders begin improving their biomechanics by addressing the relative balance and strength of the back and the front, the SBL and SFL. These lines are so fundamental – and the tendency to lose the balance between them is so strong for so many of us – that in coaching we need to revisit them regularly, even as rider and horse progress through the competition grades. Even elite riders can find improvement in their sitting and *also their horse's carriage* by improving this essential back-front balance. Their influence is not surprising, given that the latter is a demonstration of the dynamic balance within the equivalent lines (along the back and belly) in their horse.

The Superficial Back Line

The SBL is a toes-to-nose continuity (Fig. 3.1). (I refer to it in the singular, but of course there are two lines, beginning in each foot, and lying just on either side of the spine.) The SBL joins the underside of the tips of your toes to the plantar fascia under each foot. This extends behind the heel, cupping it, to become the Achilles tendon, and the muscles at the back of the calf. It continues up through the hamstrings, which insert into the back of your seat bones.

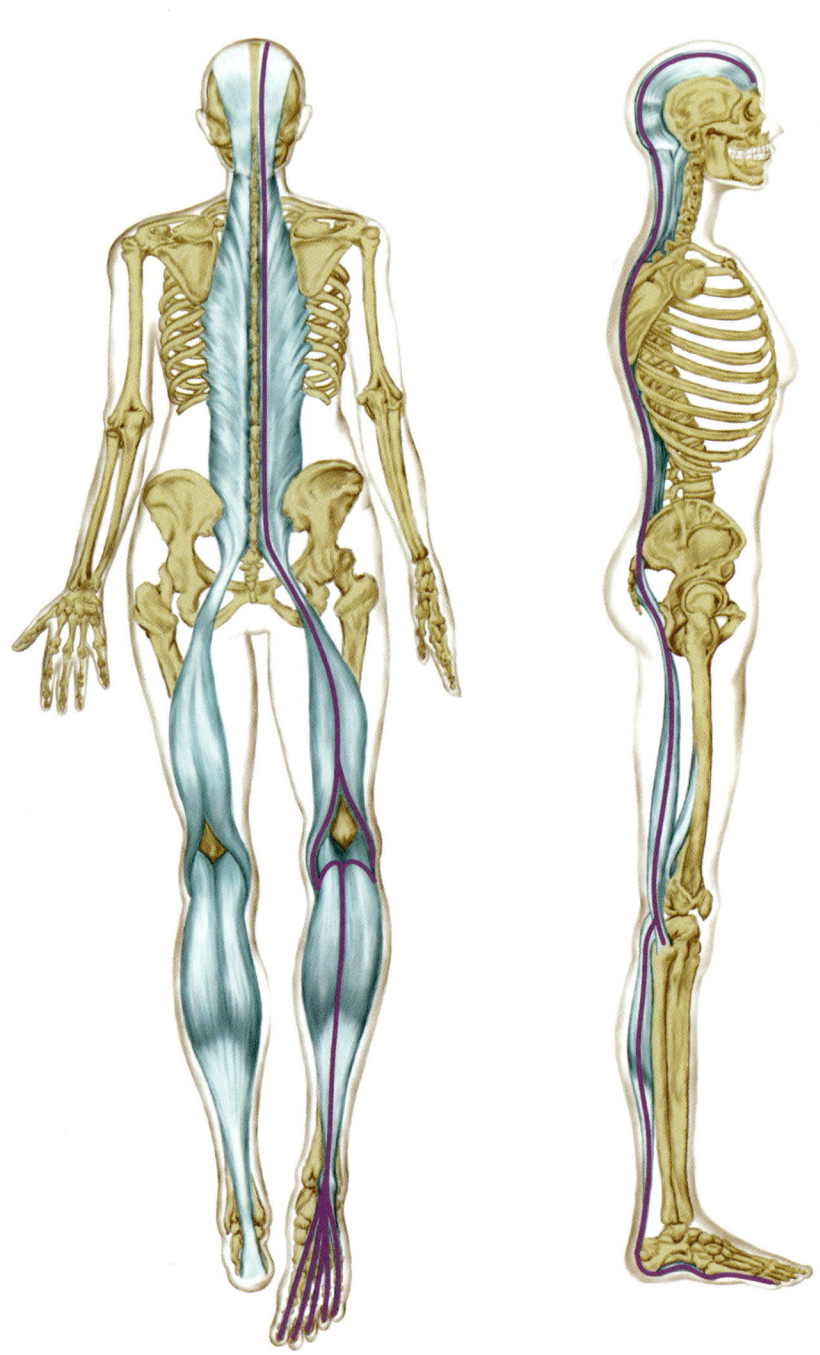

Fig. 3.1 The Superficial Back Line passes in one continuous chain of muscles and fascia from just above your eyebrows to the soles of your feet. It can be dissected out in one piece.

The SBL is an uninterrupted line, but it bends the rules somewhat behind the knee (see Fig. 3.2). The calf muscles and hamstrings connect to each other in a way that mirrors how trapeze artists hold each others' wrists. It is also like a square (reef) knot which tightens when you pull on the ends, but loosens slightly when slack. When the knees are extended straight the connection is robust; when they are bent, as in riding, Myers states that 'these two myofascial units can go their own ways, neighbouring but only loosely connected'.[1] This may explain why I know so many exasperated riders who bemoan that their calves 'go their own way'!

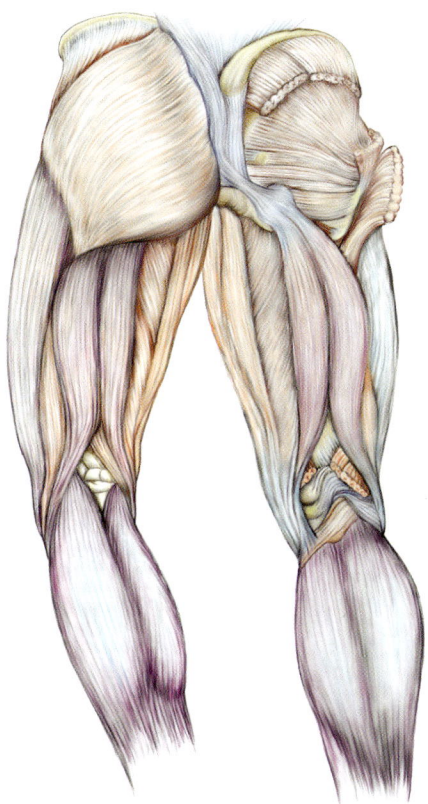

Fig. 3.2. The muscles of the calf connect to the hamstrings like trapeze artists holding each other's wrists. The right leg shows deeper muscles, the left more superficial muscles, including the gluteal muscle, which is not part of the Superficial Back Line.

Later, I will offer many ideas to make this calf-to-thigh connection more robust, making it easier to maintain a shoulder/hip/heel vertical line when riding, and to reap the benefits of sitting in a 'martial arts' posture.

From your seat bones the SBL passes under the medial side of the large gluteal muscle (i.e. the side closest to the mid-line). Only its most medial fibres could qualify for inclusion, as most of its fibres go more horizontally. A short, strong, almost bone-like ligament joins your seat bone to your tail bone (coccyx) and sacrum, and from this sacrotuberous ligament there is a continuity into the fascia that covers the sacrum. From here the line continues up your long back muscles to your head, and even on over your skull, passing under your scalp to the ridge at your eyebrows (see Fig. 3.1).

In practice you can feel the continuity from your toes to your brow line by bending forward to touch your toes, and then rolling the sole of one foot slowly and deeply over a tennis ball for about two minutes. Be thorough; include the entire surface from the pads of your toes to the front of your heel. Then bend forward again: it is very likely that the treated side of your body will reach further. Remarkably, clinical findings demonstrate that, in some cases, releasing the fascia under your feet can relieve tension headaches or low back discomfort, as it reduces constriction and increases the 'play' in the entire line of the SBL.

Your long back muscles are not directly under your skin; they are covered by more superficial muscles that do not form part of the SBL. The long back muscles themselves are long 'expresses', and underneath them there is a complex system of tiny 'local' muscles that join neighbouring vertebrae. These small local muscles have a lot to do with the mobility and overall health of your spine, and these are the muscles affected when your chiropractor or osteopath 'cracks' your spine.

There is also a set of tiny but vitally important local muscles (the suboccipitals) which join your skull to the top two vertebrae of your neck. These muscles are enormously sensitive – almost as 'smart' as your eye muscles themselves – and they have connected the movements of your eyes to your spine since your mama's mama's mama was a fish. It is hard to overestimate the importance of these muscles to easy, proper movement.

In most humans they do not work well; often they are locked really tight. In most animals, including your horse, the line of sight, the line of movement, the line of the spine, and the line of the digestive tract are the same. We humans, however, have the line of sight and movement going forward horizontally, since we shifted the spine and digestive system to a vertical orientation. Thus we tend to 'unhook' these little muscles from their proper function of coordinating between the eyes and spine.

As we are crammed into chairs and ill-fitting desks at school (for example) we gradually lose this ability to communicate unconsciously to the rest of our back musculature, separating our head from the rest of our body. In a cat, these muscles communicate so well with his eyes, inner ears and spine, that they will allow his spine to right itself even in a short fall, so that he always lands on his feet. (However, should your nervous system be this finely tuned, please arrange to land on your side and backside when you fall off; riders who land on their feet tend to break an ankle!) Loosening and reconnecting to these tiny and vital muscles is a job for a really skilled bodyworker or Alexander Teacher, and it can bring great benefits.

The SBL stretches, then, from toes to nose, and governs the balance of the primary (backward) and secondary (forward) curves of the spine, which is so essential to skilled riding. But the 'control centre' of the SBL is just below the skull at the top of your neck.

The Superficial Front Line

The SBL is counterbalanced by the SFL, and again, though we talk about it in the singular, there are right and left lines (see Fig. 3.3). The SFL begins on the tops of your toes (you can see the tendons), and passes over the front of your ankle to become the muscles and the fascial sheet in the front of your shin. The line continues over and around your kneecap into the large quadriceps muscles, whose fascia crosses the hip to the front of the pelvis.

Several muscles continue on up from here but they all go off at an angle, so we

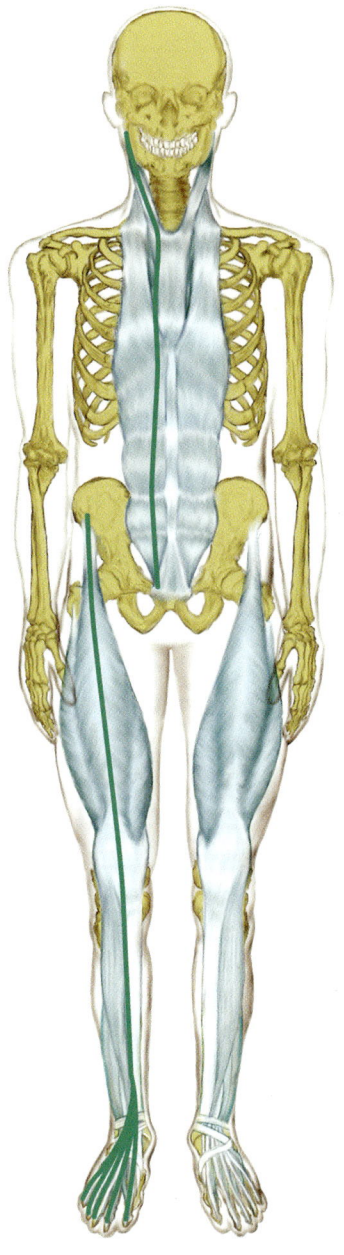
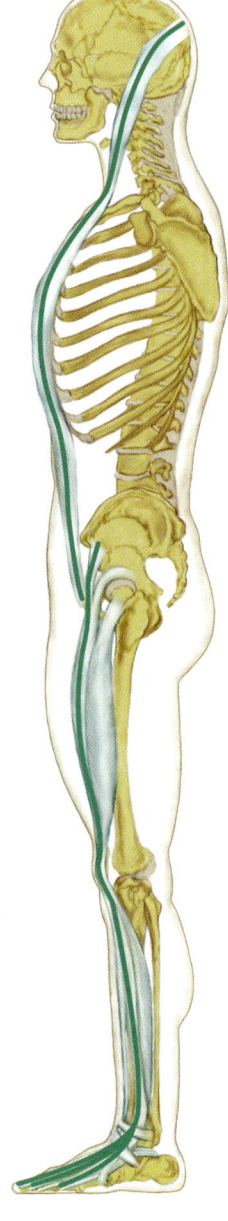

Fig. 3.3 The Superficial Front Line passes from just behind your ears to the tops of your toes. The line in the torso is separate from the line in the leg, but they operate as one for movements in the front/back plane.

will jump to the pubic bone. Strictly speaking this is a 'derailment', since it breaks this Anatomy Train into two pieces, but the upper and lower parts of the SFL function as one when we are considering movement and posture in the back-to-front (sagittal) plane. From the pubic bone the SFL extends up the abdomen and the front of your ribs and breastbone into the big muscle on each side of your neck (the sternocleidomastoid), reaching up beyond the big bone behind your ear to make a strap like a headscarf over the back of your skull.

We all start in the womb in a flexed position, with the SFL shorter than the SBL. Over our first year, we develop the strength in our SBL muscles to progressively take us from lying on our back, to lying on our front, then to crawling and kneeling, and finally to upright standing and walking. Anyone who later takes to wearing high heels will set about shortening the SBL and lengthening the SFL in their feet, calves, and thighs, with repercussions that continue on up the lines. Many people cannot flex their ankles to lower their heels, and they also have tight hamstrings and a hollow mid-back – all common postural patterns associated with a short SBL and problematic imbalance with the SFL. Correcting these imbalances promotes effective, stable riding.

The muscles of the SFL are predominantly fast-twitch muscles, reacting quickly to any shock as we instinctively crouch to protect ourselves, as in the startle response. (Prey animals, in contrast, run for their lives, except for some baby animals who are programmed to freeze.) If the SFL is reflexively short, it restricts our ability to extend the spine in a back bend. When it malfunctions it pulls the head forward, and can also pull the body into a forward, flexed position that brings the chest closer to the knees. If you crouch forward in fear when you ride and your knees come up on the saddle, this is a large part of your problem.

Many people, riders or not, come through childhood with tensional 'leftovers' of shortness in the SFL and, for the beginner rider, the simple fear of the horse, and/or of falling off, can result in chronic shortening of the muscles in the body's 'soft underbelly'. Countering this habit early and often will go a long way towards preventing back pain and neck pain in the rider, and it also spares the horse from sensing the fear that emanates from the rider via this tension.

Rather different signs of trouble could include hyperextended knees – where the knees are pushed backwards until they lock – which severely limits our adaptability, and thus can lead to low back problems. Another sign is a pelvis that is tilted so that the points of hips come too far forward and the lower back hollows – likewise a formula for distress.

A well-functioning SFL pulls the pubic bone up towards the ribs, rather than the head and ribs down towards the pubic bone. However, the latter tends to happen, either from the fear patterning described above, or simply from the effects of gravity as we progress into old age. The muscles of the SBL provide the counterpoint to this shortening, and they are predominantly slow-twitch muscles, designed for the long endurance haul of holding us upright, and limiting forward flexion. The fascia of the SBL involves some extra-heavy sheets and bands which help to make it sturdy enough for the job (for instance the Achilles tendon, the big hamstring muscles, and the big sheet of fascia across the low back).

The upshot of this difference in muscle fibres between our front and back is that, in the short term, the SFL will win any skirmishes. In other words, when we are scared, the front will win out and protect our soft squidgy bits. The SBL will fight back, hollowing our back or neck or just clamping down painfully in the mid-back to counterbalance. Whichever line wins in the longer term, we lose – unless we work on the tonal balance of the SFL and SBL.

We all have seen older people leaning down onto a walker. Their SBL has been lost for good, and whilst they were still at the stage of Fig. 1.4 page 22, they really needed the bodywork interventions that would have opened their chest and taken the strain off their aching back. (They would, however, have craved a back massage.) It is strange but anatomically true that the design of both the SBL *and* SFL contribute over time in humans to the posture of a forward head, with the SBL gradually becoming overlong and wide on the body as the SFL gets shorter.

The Superficial Back and Front Lines and 'Neutral Spine'

A good postural balance, in riding as well as life, requires tonal balance between the two myofascial complexes of the SBL and SFL. This is known by physical therapists, Pilates teachers, and personal trainers as 'neutral spine'. The spinal curves are like a set of alternating archery bows, whose 'curviness' varies between people, and 'neutral' implies that the curves are in balance with each other. If the spine were like the mast of a special boat (see Fig. 3.4) our concern in this chapter is with the relative tension in the guy-ropes behind and in front of that mast, since imbalance in their tension would bend the mast in an unhealthy way. Also, if *only one* of the front and/or back guy-ropes is extra tight or loose, there may be a rotation or a bow to one side as well.

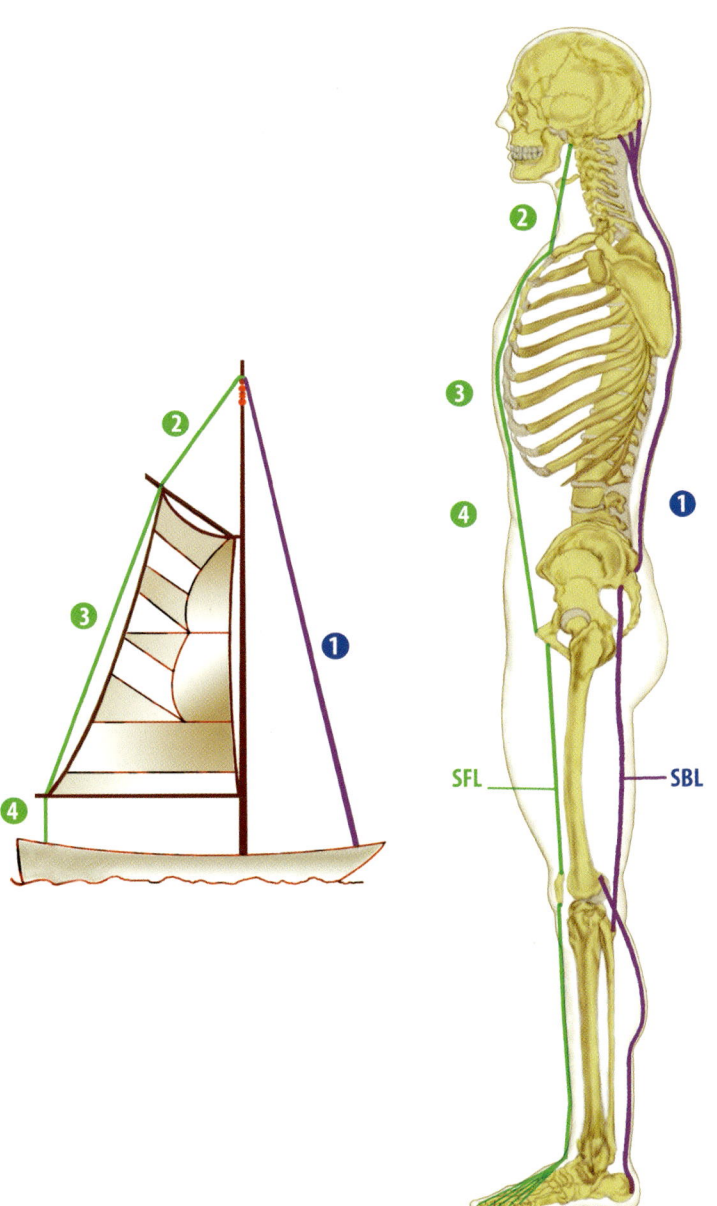

Fig. 3.4 The Superficial Back and Front Lines have a reciprocal relationship, not unlike the rigging of a sailing boat. The Superficial Back Line is designed to pull down the back from top to bottom, and the Superficial Front Line is designed to pull up the front from pelvis to neck.

Look at the photo of Heather Blitz (see Fig. 1.8(a) page 34). Her front and back are the same length, and are parallel and vertical. Also, if her horse were taken out from under her by magic, she would land on the ground on her feet. This confirms that she has a shoulder/hip/heel vertical line. The angle of her thigh bone is 50 degrees from the horizontal, which is slightly more vertical than the 45-degree angle that I recommend for less skilled riders. Her seat bones point straight down to the ground, rather than towards the horse's fore or hind feet. Heather's posture allows her to match the forces of the horse's movements so well that she is stable from top to toe. There is no 'noise' in her sitting, and no need to pull back with her hand.

Sadly, a less-than-ideal balance is more than common. If your front is too short, through anxiety or habit, your riding 'neutral' will be a cramped and tipped-forward position that is a real struggle to get out of, regardless of how many times you are told to 'sit up' or 'relax'. If you are a round-backed rider your SBL has already become too long, and a different set of adjustments will bring you into 'neutral spine', where your seat bones point down and your back and front have equal length.

Conversely, some riders try to 'grow up tall' and stick their chest out so assiduously that they train themselves into a hollow back with the SBL too short. This is the Landau, or extension reflex, which is a 'flight' reflex similar to the protective 'startle response' we discussed earlier. Many riders mistakenly assume that elongating their front whilst also leaning back is the ideal of dressage, but instead of being a pathway to optimal function it leads to back stiffness or pain. This posture of 'false confidence' (being 'up front' to a fault) actually compresses the joints that lie between each vertebra, and also the back edges of the intervertebral discs. In addition to being maladaptive for movement, both the hollow and the round back postures restrict easy breathing.

It helps to think of the torso having two major hinges: the first is the hip joint, which closes (flexes) as you lean forward, and opens (extends) as you lean back. The second hinge is the lumbar spine (between your ribs and hips) which hollows your back when you elongate your front, giving you the swayback of spinal extension. When your spine folds into flexion and your back rounds, each vertebra moves backward as you elongate your back and shorten your front.

This gives us eight permutations along with neutral, and another distortion that we will add in Chapter 8. The first four involve just one hinge: you can lean forward or lean back from the hip joints, leaving your low back in neutral; and you can either hollow or round your back whilst remaining vertical. Often, however, both joints get involved, so you flex forward at the hips and further round forward in the back, or flex forward at the hips whilst hollowing your back in a counter-compensation. Or you might extend the hips, leaning back whilst hollowing your lumbar spine, or extend the hips whilst rounding the lumbars, producing a very rounded back or a 'flat back' (which is advocated by some, but in my experience is an invitation to low back problems).

For a rider, it is important to know which of these eight options is your 'default'. Teachers also need to be able to make quick assessments of their student's patterns, to know which will become most obvious in stress situations. The ideal is shown in Fig. 3.5(a), where you are vertical and neutral, like Heather in Fig. 1.8(a) page 35.

Fig. 3.5(a) Vertical and neutral.

(b) Closing only the hinge of the hip joint to lean forward.

(c) Opening only the hinge at the hip joint to lean back. Most people cannot lean back very far without also hollowing or rounding their back.

(d) Hollowing the back without leaning forward or back.

(e) Rounding the back without leaning forward or back.

(f) Learning forward and rounding the back.

(g) Leaning forward and hollowing the back.

(h) Leaning back and hollowing the back.

(i) Leaning back and rounding the back. Most of the rounding here is around the shoulder-blades. Men can more easily round their lower back.

(j) This 'S' shape, with the head coming forward, the upper back rounding back and belly protruding forward is created by the Spiral Line, which we will meet in Chapter 8.

The muscles of the SFL and SBL become balanced, and your 'neutral spine' allows you to be upright, resilient, and adaptable.

Top-class riders stabilise their core between the SBL and SFL, especially in the mid-section of their torso. This means that the movement that keeps them 'with' the horse takes place in the hip joints instead of the lumbar spine. The myofascial lines at the back and front of the body are assisted by other stabilising lines that lie deeper within, resulting in a torso that looks completely still. However, many less-talented riders naturally adopt the 'damp noodle' technique of absorbing the movement of the trot by giving way in the middle. Not only is this the opposite of core stability, it sends a repetitive and ultimately damaging whiplash up their spine.

How to Balance the Superficial Back and Front Lines: Finding Neutral Spine

1. Since any position we habitually adopt comes to feel like 'home' we cannot trust our felt-sense to tell us what we are actually doing. So whatever level you ride at, you need feedback from a mirror, a camera, a trainer, or a good friend, to be sure that your self-diagnosis is accurate.

2. Sit on a chair, and later on horse, sideways-on to a large mirror. When you turn your head to look in the mirror be careful not to twist your torso. This is important. If a mirror is not available in your riding arena (or field!) have a friend take photographs or a video of you directly from the side. Do this from both sides – they often give a different message.
3. Identify if you are leaning forward or leaning back, and/or rounding or hollowing your back. Ask yourself (as you look at the pictures): 'If there were a string or an elastic from my collarbone to my pubic bone (SFL), and from the nape of my neck to my coccyx (SBL), would they be the same length'? If not, the lines are out of balance.
4. Double-check this by sitting on your palms, and feeling for your seat bones. Move slowly between rounding your back and hollowing your back until you think you have found the place where your seat bones point straight down onto your hands. If they had flashlights on them, would these point down to the ground, forward towards the horse's forelegs or back towards his hind legs? Double-check your perceptions by looking in a mirror as above. Familiarise yourself with the exact point of pressure on your seat bones, and re-create the pressure on that point when you are in the saddle. When you are vertical, with your back and front at equal lengths, your seat bones should point down.

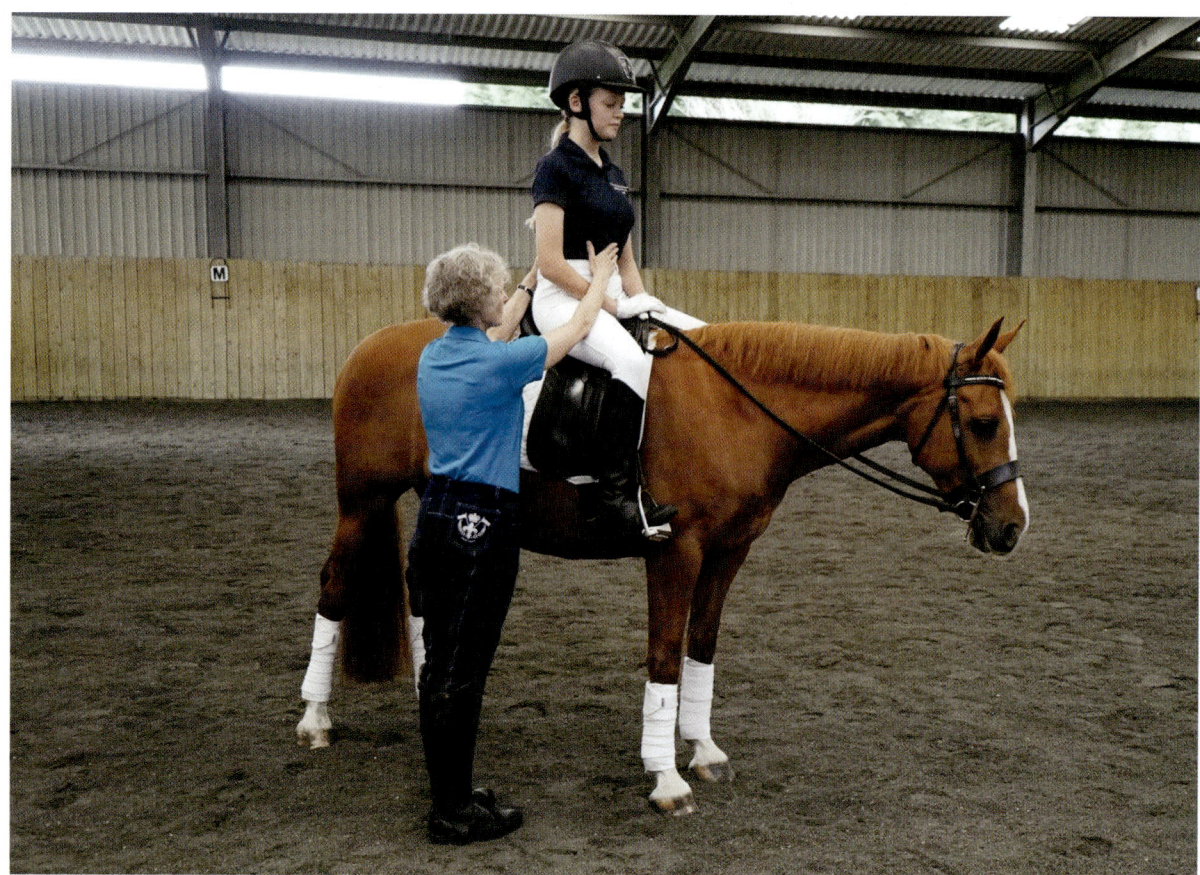

Fig. 3.6(a) Begin by finding neutral spine with the legs up over the saddle. When the seat bones point straight down, the rider's front and back are the same length. The coach brings down the leg and makes small circles in each direction with the knee, feeling how freely the thigh can move in the hip socket.

5. On your horse, alter the length of your stirrup leathers if necessary so that the outer centre line of your thigh is at 45 degrees – halfway between horizontal and vertical. If you are a skilled upper-level rider, your thigh could be slightly more vertical, with a 50-degree angle. Then ask yourself, 'If the horse suddenly disappeared out from under me, how would I land on the riding arena?' If the answer is not 'On my feet' but rather 'On my bum' or 'On my nose', then you do not have a shoulder/hip/heel vertical line. This indicates that you still have some postural work to do to achieve equal tension in each line.

6. Many riders, in their attempt to get their heels down, extend their knees and push their feet too far forward. Ideally, a coach would help you reorganise your legs (see Fig. 3.6), but in the absence of outside help, grab the back of each thigh in turn and pull it outwards to rotate the front of your thigh in towards the saddle. This manoeuvre makes it easier to bring your feet back under you.

7. Turn your head to look in the mirror. Are your front and back parallel and vertical? Are you in a martial arts posture (known, appropriately enough, as the horse stance)? Would you land on your feet if the horse were taken out from under you? This 'seat bones pointing down, neutral spine' way of sitting is a good idea for *all* sitting, so take a second photograph and check your kinaesthetic perceptions of how you are sitting now against how you look in the photograph.

(b) She then brings the leg out and back before placing it on the saddle from the back to the front, so that the flesh of the inner thigh is pulled around to the back.

(c) She checks that the stirrup is under the ball of the foot with the foot resting lightly in it, to give the rider a shoulder/hip/heel vertical line.

(d) She double-checks that the rider's torso has remained neutral.

8. Changes in your body will always feel bigger than they look, and you may well feel more weird than you could ever have imagined! Realise that when you lose a filling or get a sore on your tongue, it feels huge but looks tiny in a mirror. So it is in riding, and this discrepancy stops many people from making the changes they actually need – the comfort of habit is a very strong force. Seeing is believing; comparing your visual and kinaesthetic experience will encourage you to be bold and persistent in the changes you make. Learn to embrace new weird feelings until they become your new 'home' – it will pay you handsomely.
9. Find some words to describe the changes you feel. They can be any words that work for you, and they can express how much you do not like this change, or how much it does not feel 'right'. Remember that your old way of doing it was not '*right*', it was just *familiar*, and this new, weird way of feeling is not '*wrong*' it is just *unfamiliar*.
10. Once you are sure that you are vertical and neutral, with your heel under your hip, ask your friend to give you a resistance on your upper chest (see Fig. 3.7). Do not let her push you back, but also do not overpower her so that you lean forward. As you ride, imagine that the faster you go, the more likely you are to get pushed back by an increasingly strong wind. Resist that temptation, and realise that this resistance is activating the SFL, potentially resulting in a shortening from your collarbone to your pubic bone.

Fig. 3.7 The instruction to 'resist my push' gets the rider to 'switch on' the muscles of the Superficial Front Line, and becomes a useful reminder when she is in motion.

11. Your friend could also give you a resistance on your upper back (see Fig. 3.8). This is harder to resist, and you might find yourself lifting off your seat bones and/or toppling forward. If necessary, have your friend begin with a smaller resistance on the back of your waist, and then work further up your back as you train yourself to become more resilient.

Fig. 3.8 It is more difficult than you might expect to be unmoved by a resistance on your upper back. Begin with a resistance given lower down, being sure that you do not lift your backside or curl your toes.

Breathing and Bearing Down – an Introduction to Your Core

Whilst breathing involves every line and is fundamentally a 'core' function, we will start here with breathing as one of the fundamental skills you need to build as you overhaul your riding. Since good breathing is a life skill that can even help you live longer, start now!

Alongside good breathing, the strength that I call 'bearing down' gives you the torso strength to begin matching the forces of the horse's movement. It is best to learn this with the help of a balloon. If you do not have one, you can use a drinking straw, or purse your lips as if breathing out through a drinking straw. Whilst this works well enough for starters, it will be worth your while to buy some balloons!

If you are someone who has never been able to blow up a balloon (about one in ten people), practise with small soda or cocktail straws to develop your strength. Also, it is easier to blow up a balloon that a friend has already blown up. Do not give up; you can gradually develop this hidden strength – and believe it or not this skill underlies centred riding.

1. Normally when you blow up a balloon, you pinch it off and take it out of your mouth as you breathe in. In this version you are going to keep the balloon in your mouth throughout the in-breath, without pinching it off or sticking your tongue in the hole! This means that you have to breathe in (through your nose, of course) *without letting the balloon empty itself into you.*
2. Before you blow up your balloon, you need to notice what parts of your body move most with the breath. So we will first take an inventory of your normal breathing. Sit quietly, in neutral spine, and take a few breaths. What part of your torso moves most with the breath? Is it your upper chest, lower down in your ribs, or your belly? Do your ribs *lift* as you breathe in, or do they expand without lifting, so that your ribcage increases its circumference and diameter? Is the movement of your ribs primarily or entirely in your front? Do your sides and back expand? Does your belly also expand?
3. Once you are clear on this, blow up the balloon (without pinching it off) and after you have done so, hold it whilst you keep breathing in the same way. How is this different? Most people find that they are breathing much more into their back and sides, and that they are no longer lifting their ribs. If your pattern was to breathe into your upper chest, you might feel as if your ribs have been pushed down and together near your sternum.
4. What happens to your ribs on the expulsion part of the breath? On a normal out-breath your ribs deflate; but as you breathe into the balloon you want your ribs to *stay expanded*. Not everyone manages this on their first attempt, and you may have felt your ribs deflate instead.
5. Let the air out of the balloon and begin again, really noticing your body, and aiming to keep your ribs expanded on the out-breath. Also notice what has happened in your belly. You will find that you are holding it firmly. What you have actually done is to *pull your tummy muscles in to make a wall while your guts (which cannot compress much) push out against that wall.*

In medical parlance, this bearing down is called the Valsalva Manoeuvre, and it increases your abdominal pressure, which in turn tensions the surrounding fascial net. You naturally increase your internal pressure when you cough, giggle, vomit, clear your throat, play a clarinet, sneeze, defecate, blow your nose, bear a child and (I hope) when you sweep your yard or barn. I expect you have seen and heard some of the world's best tennis players grunt to increase the power in their core, giving them more power in their shots. In martial arts, this expulsive sound is called the 'Ki-Ai', and imparts more power to kicks and punches.

When I discovered this, I called it bearing down. The concept of core strength had not yet been invented, and sadly for my pocketbook I did not put it in those terms; but bearing down – gathering of your belly strength – is your first introduction to core strength as accessed by riders. The terms 'bear out', 'bear forward' 'bear against the wall' are also viable, but the best description is: you pull your stomach in to make a wall, and then push your guts against that wall. Do not push your weight harder down into the saddle as you do this. Continue to use these exercises to locate your core in your own kinaesthetic perception.

1. Practise bearing down without the balloon. As you do this put one hand just below your sternum, and the other on your bikini line (see Fig. 3.9(a)). (For male riders and anatomy geeks this is the inferior portion of the transversus abdominis and internal oblique. Some male riders have suggested that I call it the 'speedo line' - but I hope they will forgive me if we just use the term 'bikini line'!) Clear your throat, and feel your muscles firm up more. Make the sounds 'Shhh' and 'Pssht'; most people find that 'Shhh' makes their solar plexus firm up (just below their sternum, and key weak place for round-backed riders). 'Pssht' usually makes the bikini line area firm up. This is a particularly important exercise for riders who have 'grown tall' and sucked in their stomach in the process.

Fig. 3.9 Learn to bear down by giving yourself resistances. (a) Make your fingers into a claw, with one hand beneath your sternum, and one on your bikini line. Make the sounds 'Shhh' and 'Pssht'.

(b) Put your hands on your back and front, and clear your throat.

(c) Put your hands on your sides, and make the sound 'Ssss'.

2. If this is not clear to you, try a hearty 'Grrrr!' or 'Brumm brumm', as if you were revving your internal engines, which indeed you are. If your 'Grrr!' and/or your 'Brumm' are high-pitched and a bit pathetic, keep working on them until you sound fierce and can feel your voice vibrating your lower belly! Become convincing; your horse has to believe in the power that you can develop within your core.
3. Clear your throat again. Put your hands on your sides between your ribs and hips, and feel your sides push out against this pressure (see Fig 3.9(c)). 'Ssss' may be the sound that helps the most. Keep experimenting to find the noises that make the most difference. (Kids have great fun with this, but adults often baulk at the idea. But given that it does not embarrass elite tennis players, why should it embarrass you?)
4. Try bearing down with one hand below your sternum, and the other on your back, putting your thumb and fingers each side of your spine just beneath your ribs. Feel your back muscles firm up against your hands. It is important that your back fills out as you bear down, and this is often difficult for hollow-backed riders (see Fig. 3.9(b)).
5. Practise whilst driving your car, sitting on trains and planes, pushing the supermarket trolley, and, of course, riding your horse! Keep looking for your weakest place and gather it into the fold; your aim is to make bearing down and diaphragmatic breathing into a way of life.
6. Blow up another balloon, and really feel the involvement of your back and sides, as well as your abdominal muscles. Then take the balloon out of your mouth, continue breathing in the same way, and notice too what has happened inside your head. I think you will find that it is very quiet in there, with little or no internal chatter, and a sense of peace and calm. This may enable to you to discover the difference between *trying*, and *noticing*.
7. Notice too that you are using your unfocused or peripheral vision, which enables you to see everything without looking directly at anything. Enjoy this state; it is part of the mindset and brainwave pattern that puts all elite sportsmen and women 'in the zone'. Here they appear to perform effortlessly, and to have plenty of time.

All these exercises not only strengthen your belly core; they also stimulate your 'vagus brake'. The vagus nerve goes to your various organs, and it switches on the hormones of 'rest, repair and digest' as opposed to 'fight or flight'. If you are a somewhat nervous rider, bearing down and breathing like this will make you feel safer in two ways – by increasing your stability and, paradoxically, since it involves apparent effort, relaxing you as well.

Realise that *every time you find a new way to increase that pressure in your insides, you can expect it to feel stressful at first*. Many riders fall at the first hurdle when they refuse to believe that skilled riding requires this. Persevere: the benefits will soon become apparent to you, and the feeling of effort will eventually fall away, potentially leaving you with enough 'amnesia' to say 'I'm just sitting'!

How Different Riders Can Find Neutral Spine

Finding Neutral Spine as a Hollow-backed Rider

If you began your realignment from a hollow-backed position, any of the following reminders could be the one that 'does it for you'. I have worked down the body from the top, but your aim is to find the most effective cues for you – which may change over time.

- Drop your chin. Some riders begin the process of drawing their chest up and hollowing their back by lifting their chin and hollowing their neck. Notice whether you yourself are 'on the bit' or above it!
- Bring your collarbone over your sternum, rather than behind it. Check this by putting your thumb into the indentation in the centre of your collarbone, and curling your fingers out of the way so that your knuckles are against the central bone of your ribcage. Reach your little finger down towards your sternum (see Fig. 3.10). Experiment with arranging your ribcage so that your thumb is behind your sternum, then over your sternum, and in front of your sternum. When riding, drop your ribs so that your thumb stays vertically over your little finger.
- Drop your chest – think of having 'low beams'. Where on the horse's head does your chest need to aim for you to remain in neutral spine? It could be the poll, browband, eyes, or even the bit rings (depending on your size and his head-carriage).
- If you are accustomed to your hollow-backed riding posture, trying on these changes may make you feel slouched, scrunched, or slumped. They may also make your shoulders feel rounded, and you need a way to check whether this is really the case. Keep relying on the objective feedback of the mirror and your friend; your felt-sense of any difference will be exaggerated.
- Notice whether your armpits feel more closed at the front or the back. If they are more closed at the front, you need to close them more at the back (see Fig. 3.11 page 68). Bring your shoulder-blades down and together without lifting your chest. If, as a result of these changes, you find yourself looking at the horse's withers, let this be for the moment. If you bring your head back up you are highly likely to lift your chest back up with it. We will adjust your head position more effectively in Chapter 8 using the Spiral Lines.
- Imagine being punched in the stomach, and rounding your waistband back.
- Make an extra crease in your front between your sternum and waistband.
- As you learn to bear down, make sure that your 'push' also goes into your lower mid-back. You will almost certainly have to feel round-backed in order to not be hollow-backed.
- Think of lifting your pubic bone, so that you make a few horizontal creases between your pubic bone and your bikini line.
- If your pelvis were a bowl full of liquid, would the liquid spill out of the front of the bowl? Can you lift the front of the bowl so that the liquid stays level inside

Fig. 3.10(a) Put one thumb in the centre of your collarbone, with the backs of your fingers against your chest, and your little finger reaching down to touch your sternum. Adjust your torso so that your thumb is behind your little finger as in (b), over it as in (c), and then ahead of it. Find and maintain the neutral where your thumb is directly over your little finger as in (c).

(a)

(b)

(c)

it and your seat bones point straight down? Train yourself to notice every time they peel up off the saddle, until you can keep the 'back edge' of your 'torso-box' down all of the time.

Finding Neutral Spine as a Round-backed Rider

If you began your realignment from a round-backed position there are also many ways of making the desired change, and any of the following reminders might be the one that 'does it for you'.

- If your neck stiffens up as you ride (this is especially likely in sitting trot) think of your head like a ball balanced on a seal's nose, and make very tiny movements which will help to keep your head and neck free.

(a)

Fig. 3.11 The armpits are closed at the front in (a) and the back in (b). Realise that this change does not involve lifting or lowering the chest.

- Lift your chest by bringing it more through between your elbows.
- Notice whether your armpits are more closed at the front or more closed at the back. Close them more at the back (see Fig. 3.11).
- Flatten your upper back. Draw your shoulder-blades down and together. Think of them slotting into the back pockets of your breeches.
- Pull down your 'lats' (latissimus dorsi, the big muscle that covers much of your back (see Fig. 3.12), which we will meet again in Chapter 6 on the Functional Lines). If you reach one hand across your front and grab just under the back of your armpit, your fingers will be around your lats. Pull down your shoulder and elbow and feel for the big tendon which bulks out against your hand.
- Advance your sternum, and think of it like the prow of a ship, carving its way through the water. You can do this with your waistband and/or your belly button as well.

Fig. 3.12 Reach one arm across your chest and grab the back of your armpit. Pull down your shoulder and elbow, and feel the edge of your lats bulk out. Also find the tendon that inserts the lats into your arm bone.

Fig. 3.13 Sitting on a kneeling stool, that is good for your posture, spreading weight down through your thigh.

- Bearing down and repeatedly making the sound 'Shh, Shh, Shh' may help you lift whatever part of your front is most collapsed.
- Advance the front points of your pelvis and imagine that you can stick them out through your breeches. Notice how this puts more weight onto your inside thighs, so that you 'kneel more', with your knees reaching further down past the saddle.
- Keep your pubic bone down, and keep your weight *in front of* your seat bones, not back towards your tail bone. You will have to feel hollow-backed in order to not be round-backed.
- Keep your thighs against the saddle and 'kneel' as if sitting on the Norwegian stool of Fig. 3.13. If you fall back into a round-backed position, they will rotate outwards as your knees come up. Your thighs may well be 'the first domino to fall', starting a cascade of changes that rounds your back and topples you backwards. Stop frequently and rotate your thighs inwards, to mimic Fig. 3.14.

Fig. 3.14 In photographs of elite riders riding towards the camera, you will always see the inside thigh on the saddle.

- 'Tight is light'. Many round-backed riders sit too heavily in the saddle, as they are not supporting their bodyweight well. Think of your pelvis wrapped tightly in a corset; do not let it 'splodge' onto the saddle.
- Imagine being hung in a harness that goes under your pelvic floor, like a climbing harness. Use this to help you support your bodyweight, as you also think of putting more of that weight down through your thighs. You may need to feel that up to 80 per cent of your bodyweight is being taken by your thighs.
- Think of how the male dancer picks up the female dancer in ice-skating, and imagine being picked up like this, perhaps from the outsides of your pelvis. They provide a wonderful example of tensegrity in motion!

Checking and Strengthening Neutral Spine

Whilst riding in walk, your aim is to maintain neutral spine, and to take any wiggles and jiggles or shoves and pushes out of your backside and torso. Notice, and have your friend notice, which layer of your torso moves the most. It could be your shoulders, your lumbar spine, or your pelvis. Your aim is to move only in your seat bones, which 'walk along' making small movements that follow the seam lines of the saddle, with no shove and no push. The rest of your torso should be very close to still, as if you were wearing a pelvis-to-armpits corset. This is easier said than done, but bearing down is a big part of the secret.

Some riders baulk initially at the feeling of being more stable, thinking that it must be making them 'stiff'. Sometimes they complain that they are 'rigid', 'set in concrete' or 'like a statue'. One rider christened herself 'marble woman' when she really got it! The underlying thought of these riders is, 'Surely I should be more supple and go with the horse's movement?' Becoming more still challenges their mindset as well as their body. The amount of movement that they naturally have in their torso (and that they may well have worked hard to cultivate) is overkill. It classes as 'noise'.

Fig. 3.15 Push and pull on the back of a chair, and also the saddle. Feel your front firm up when you pull, and your back firm up when you push. The latter is harder for most people, and your front may firm up before your back. If you are persistent, you will discover how to access your back.

Contrary to their perceptions, they were all becoming better stabilised riders, more able to keep their centre of gravity in place over the horse's centre of gravity, without it falling backward or jiggling around. The still rider becomes much more able to match the forces of the horse's movement and to give her hands forward. Moreover, this is not just about learning how to 'use your back', but also how to 'use your front', and ultimately how to access your core.

Once you can maintain neutral, and it feels less weird, you can use the following exercise to strengthen the SFL and SBL, and also some deeper muscles in your back. Do it regularly at work every day – no one will notice!

1. Sit in a strong dining chair. Make sure you are in neutral spine. Put your hands behind you and pull on the chair's back. Feel the SFL firm up.
2. Conversely, you can strengthen the SBL by making fists and pushing back on the chair at the level of your mid-back. This is a little trickier, and your front may firm up before your back, so be persistent. Be sure that you are not leaning back, and that you push *back* and not down.
3. Whilst riding at halt or walk, you can strengthen the lines if you drop the reins and push and pull on the cantle. This can be done with just one hand targeting that same side of the body. If one side is wobblier, do the exercise first with that hand, then with the other hand, then again to target the wobblier side (see Fig. 3.15).
4. Whilst sitting in neutral spine, imagine a bolt low down in your pelvis on each side, running from back to front. The head of the bolt is on your back, and the nut on your front. Are both bolts horizontal and level? Is one wobblier than the other? If so, can you mentally tighten the nut on that side? As you do rising trot, do they go 'up level, down level', and if not, can you fix this?

The Superficial Back and Front Lines in the Thighs and Calves

Just as the SBL and SFL create a dynamic balance between the front and back of your torso, so they also balance the front and back of your thighs (the quads and hamstrings), the front and back of your calves, and also the tops and bottoms of your feet. Differences in tension between the SFL and SBL on left and right side make it highly likely that your legs are not a pair, with imbalance *within* each leg and also *between* each leg. Other lines get involved in this too, and your feet can be pulled every which way by the five lines that extend down to them. Is it so surprising, then, that even Charlotte Dujardin admits to having a 'funny leg'?

If you are lucky, you will have one leg that has a pretty good balance between the SFL and SBL, which makes it relatively easy to maintain a shoulder/hip/heel vertical line on that side; but the other leg may have its own ideas. The 'funny leg' (I have even known riders call it 'the alien leg'!) may have a shorter SFL and a longer SBL, which puts the heel forward and down, often far ahead of the ideal shoulder/hip/

heel vertical line. Or the 'funny leg' may have a shorter SBL and a longer SFL, so the foot is pulled back and on tiptoe. The rider with a short SBL on one leg may even curl her toes on that side.

Unfortunately, riders who are actively learning to ride in balance often discover that they have a different kind of 'funny' on each leg: one with a short SFL and long SBL (heel down and forward), and with one a short SBL and long SFL (heel back and up). This is usually accompanied by a twist to one side or the other in the pelvis. Only a small percentage of riders default to heel back and up on both legs, whilst riders who are determinedly pushing their heels down appear (on a superficial glance at least) to have trained the same problem into each leg.

Diagnose yourself, again with help from a friend with a camera. Take photographs or a video from the inside of a 20m circle in each direction. Have your friend also record what happens to each leg when it is on the outside, since a leg that behaves itself when on the inside of the circle might not behave on the outside, and vice versa.

All of the suggestions below presuppose that your stirrup leather length puts your thigh bone at 45 degrees, and also that your feet are resting lightly in the stirrups, rather than pressing down on them. Adjust your stirrup leathers if necessary.

1. If you have previously pushed in your stirrups in your attempt to 'get your heels down', be prepared to slay this sacred cow! If you struggle to make this change, imagine someone's fingers under your foot, between it and the stirrup, and keep reminding yourself not to squash them!
2. Realise that pushing into your stirrups causes your seat bones to come up off the saddle as you straighten your knee and hip joints. Inevitably you brace through both the SFL and the SBL, losing resilience and adaptability.
3. Demonstrate this to yourself by pushing down into one foot one whilst sitting in a chair: you will notice the seat bone on that same side lifting. By Newton's third law of motion, your *push down* has generated an equal and opposite *push up*. This is not what you want when riding!
4. Keep your foot light in the stirrup with your seat bones pointing down, and drop your breastbone down towards your hips so that you keep 'neural spine'. The advice to 'grow tall, stretch your leg down and push your heel down' is likely to get you into serious trouble!

To Elongate the Superficial Front Line and Shorten the Superficial Back Line

In other words, to get your knee down more and your foot back more, use the following images. Realise that you may need these corrections on one leg, on both, or possibly not at all. However, most riders have at least one heel whose 'default position' is too far forward.

1. Think of sitting in the saddle as if you were kneeling on one of those Norwegian stools (see Fig. 3.13 page 70). Do not treat the saddle like an armchair. Your

inside thigh needs to be part of your sitting surface, so think of kneeling, with your inside thigh on the saddle and bearing some weight.

2. If you have been taught to keep your thighs and knees off the saddle, slay another sacred cow! Look at Fig. 3.14 page 71, and notice how the thigh also lies on the saddle in other photographs of elite riders. This is one of the most significant areas where expert advice and the facts on the saddle are out of synch.

3. What percentage of your weight would be taken in your thigh and what percentage would be taken on your backside? Start with 80 per cent on your thighs, and only 20 per cent on your backside – probably the reverse of what you thought! But realise that when you have most of your weight concentrated in the centre of the horse's back, he tends to shorten his SBL and sag in his SFL – in other words, this encourages him to hollow his back.

4. Find the bony knobble just above the inside of your knee, and hook it to an imaginary bar that passes through the horse's ribcage. Do your knees stay still on that bar, and can you use this thought to stop them moving either up and down, or on and off the saddle? Is one knee more stable than the other? The bar is not an elastic that pulls your knees together – it just gives them a stable place to be. If you are tempted to grip with your knees, think of the bar holding them apart. If your knees come up easily, think of the bar being lower down and further back on the saddle than your knees and your short front line would like it to be.

5. Would the bar through your knees be level? Which side needs adjusting to bring both thigh bones to that 45-degree angle?

6. Imagine that, if the line of the front of your thigh continued downwards, it would meet the ground just in front of your horse's forefoot: keep 'kneeling' enough to maintain this angle.

7. When you are sitting with a shoulder/hip/heel vertical line, the angle behind your knee should be about 90 degrees. When the SFL is short the angle will tend to get bigger than this as your foot goes forward. Think of holding a tennis ball in that angle behind your knee. Alternatively, imagine your knee supported by a jump cup placed just underneath it, or think of the upper part of your calf supported by an imaginary kneeler, as in church.

8. Put your fingers behind your knee and feel for two tendons on the inside, and one on the outside (see Fig. 3.2 page 49). The two on the inside may feel like one until you poke around and get a fingertip between them. Can you make all of those tendons stick out? If the tendon on the outside of your knee is harder to access, turn your toe out. Can you keep the tendon sticking out as you then bring your toe forward? If the inside tendons are harder, turn your toe in. Practise this off-horse as well as on, so that you learn to stick out those tendons at will. This helps to tone and shorten the SBL – feel how it makes your hamstrings bulk out.

9. To help you shorten the SBL in the back of your calf, push it back into a resistance (see Fig. 3.16 overleaf). Notice how this helps you to access the tendons behind your knee. A resistance band like Theraband works particularly well for this. Practise off-horse by looping it around a table leg and then around your heel,

Fig. 3.16(a) Pushing the calf back into a resistance helps to strengthen the Superficial Back Line in the leg. Where do you feel muscles firm up as you do this?

(b) Using a rubber band to give a resistance gives the rider a much clearer feeling, and allows the addition of 'toes up'. The foot must be out of the stirrup. Be aware of safety issues if you do this.

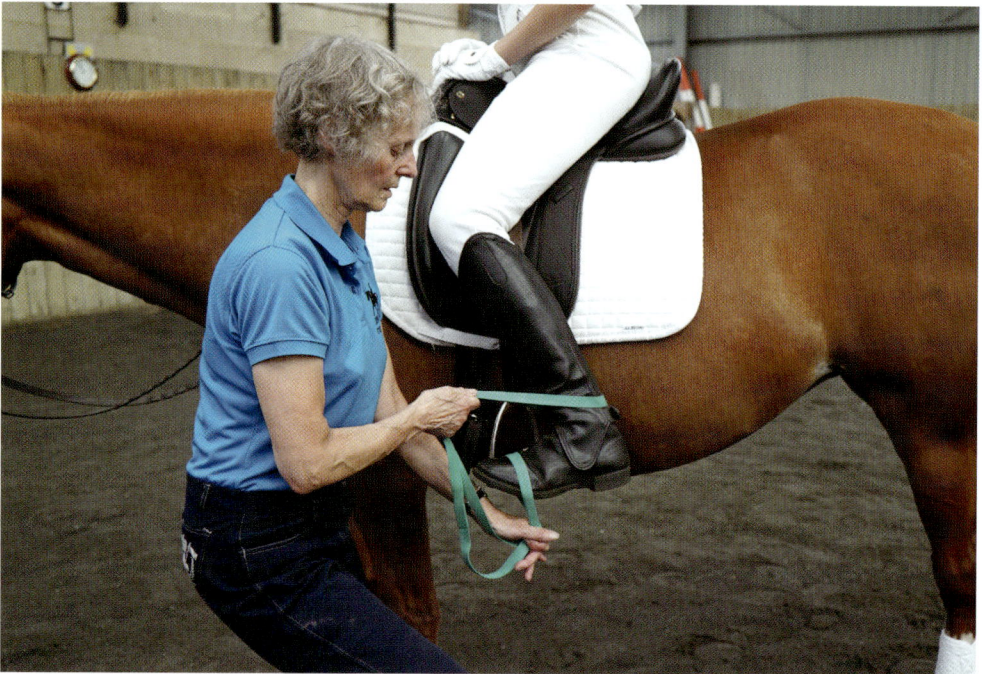

and exert a fairly strong resistance for just a few minutes until the hamstrings feel definitely tired. Doing that even two or three times a week will quickly make a difference in your hamstring isometric strength – which is what you need in the saddle.

10. Encasing the lower leg in a resistance band as shown in Fig. 3.17 makes the calf feel much more contained, and helps many riders to hold it in place. Afterwards, think of the stirrup coming *up under your foot* just as the band supports your foot from underneath.

Fig. 3.17 Encasing the calf in a rubber band helps riders to feel and control it. The difference can be dramatic! Be aware of safely issues as you do this, and only do so at halt.

11. If you are failing in all of your attempts to elongate your quads and get your knee (or knees) to go down more, think instead of *shortening your hamstring muscles* just in front of your seat bone. Find your seat bone with your fingers, and then place your fingertips a bit down from there, along a line beneath your thigh bone (see Fig. 3.18). Attempt to firm up the muscle under your fingers, so that it bulks out against them. Practise this, and then hold it in movement, thinking of 'kneeling down your hamstrings'.

Fig. 3.18 The part of the thigh under the arrow needs to bulk out for the rider to keep her foot back in place. Feel this with your own hand.

12. This last strategy uses the SFL in your torso to change the SFL in your thigh and calf, and is a brilliant demonstration of how they act as one line of pull (see Fig. 3.19). I am sure you are familiar with the idea of your abdominal 'six-pack'. Everybody has one, whether visible or not. Just under your sternum is what I call your 'two-pack line', which marks the bottom of the first obvious 'pack'. Imagine that your leg is like a long pendulum that hangs from here. Have someone give you a resistance by placing her fingertips along that line. Push into this resistance, and think of holding your leg on the backswing of the pendulum, with help from 'kneeling down your hamstring' if necessary. You will probably only need this on one leg, and your two-pack line may be much 'soggier' on this side of your torso-box. Really feel it *advance and firm up* as the SBL becomes more toned all the way to your toes, which will tend to lift. Remember, this change will have to feel bigger than it looks! You may need to see a video or a photograph before you believe it. Alternatively, if you have been failing to fix this leg for years, you will feel that you have finally found the answer!

Fig. 3.19(a) My hand is on the 'two-pack' line. Resist it forward, and think of the leg hanging from it, like a long pendulum that must stay on the backswing. Also think of 'kneeling down your hamstring' and of pushing your heel back into a resistance.

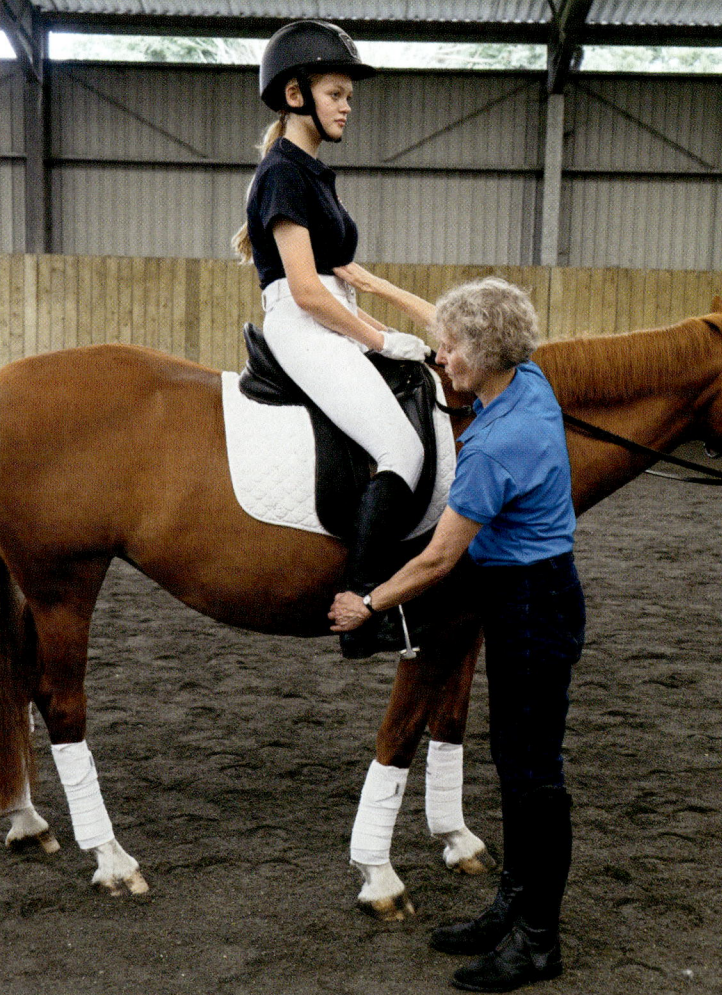

(b) When the 'two-pack' line collapses backwards, the rider's foot comes forwards, and her torso rotates to this side.

To Elongate the Superficial Back Line and Shorten the Superficial Front Line

In other words, to get your knee more up and out in front of you, with your foot more forward, and your heel more down. Realise that you may need these corrections on one leg, on both, or possibly not at all (if both legs go too far forward).

1. First realise that even though conventional instruction often implies that rider's knees should always be more down, one or both of them do sometimes need to come more *up*. If you tip forward easily, and/or you find yourself on tiptoe, and/or you give leg aids with your lower leg coming backwards, suspect that you need this fix. Which leg is the bigger or only culprit?
2. Check that both knees stay on the imaginary bar (see page 75) with the bar level. *Lift* the knee (or knees) that would slip down off the bar thus making your thigh (or thighs) too vertical.
3. Have someone put her hand on the front of your thigh just above your knee. Push your knee up into this resistance. At the same time, put your fingers in the hip crease between your torso and thigh on that side. Feel for a strong tendon (actually there are two, but treat them as one) (see Fig. 3.20). These tendons should stick up as your knee pushes up. Keep this feeling of 'front tendons up, knee up' as you ride.

Fig. 3.20 To keep the knee more up you may need to keep your front tendons more up. Find them by putting your fingertips in the crease between your torso and thigh, and lifting your knee. Then push your knee up against a resistance. Feel the tendons stick up, and notice how your thigh muscles engage.

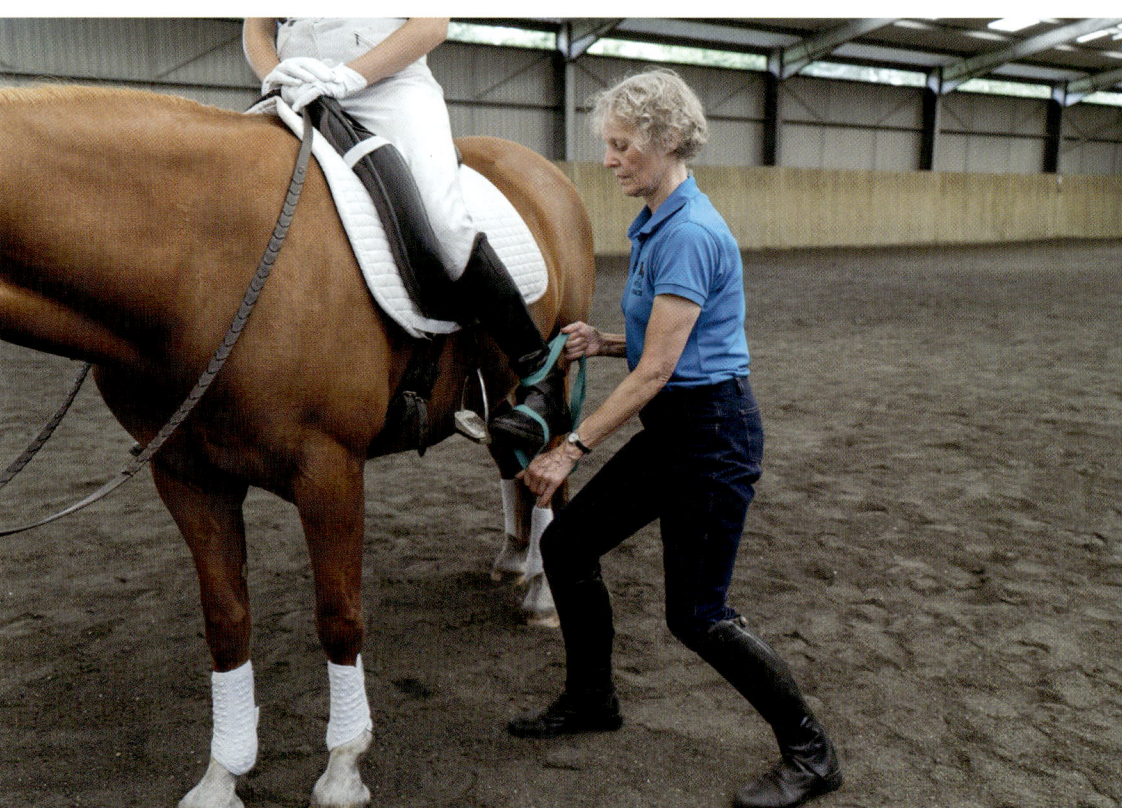

Fig. 3.21 If you need to keep your calf more forward, push it or your toe forward into a resistance. This is particularly useful if you jump.

Fig. 3.22 When this resistance is done with a band, the foot must be out of the stirrup. The band enables us to add 'toes up'. Be aware of safety issues as you do this.

4. Think of your thigh bone being more out in front of you, and being made of solid bone. If the bone feels 'mushy' or rubbery, think of your thigh muscles hugging the bone harder, as if they form a firm elastic bandage. Do you have equally 'bony bones' in each thigh?
5. Have someone give you a resistance that you push the lower part of your calf forward into. If you are using a resistance band, ask your coach to also loop the end of the band above the line on your boot when your toes begin. Resist your toes *up* at the same time as you resist your calf forward. This is the feeling that conventional instruction is attempting to invoke by saying 'heels down', but 'toes up' is a much better way to tone the SFL. This is a particularly good way to stabilise the lower leg for jumping(see Figs. 3.21, 3.22).

Toes and Tendons

As we have seen, imbalances between the SFL and SBL show up everywhere – including right down in your feet. This next exercise, which tones the SFL, begins to solve the problems of curling toes, disorganised feet, and flapping, out-of-control calves.

1. Start by lifting your toes and spreading them.
2. If, as you do this, you put your hands around your ankle, you will feel the five tendons at the front (part of the SFL) stick out against your fingers, and also feel the Achilles tendon at the back (part of the SBL) stick out against them.
3. If you then put your fingers just outside your tibia, which is the bone you can feel at the front of your calf, you will feel the muscle there bulk out.
4. The back of the calf bulks out too and the muscles on the inside of it form a ridge. If your calf muscles are bulked out like this when you give a leg aid (kick), you will more easily be able to make it like a karate chop: a small, quick but powerful movement that rebounds off the horse's side. Aim to keep your toes up and spread all the time as you ride.
5. As you sit in a hard chair, place one foot on top of the other, and then attempt to lift your toes and the ball of your bottom foot *up* against the resistance of the top one. How far up your SFL do you feel the effects of this? Potentially the line will become toned right up to your chest and neck. Alternate toning it and letting go; which one enables your SFL to 'play a note?' Practise until you have a clear sense of what it feels like to engage it.

CHAPTER 3 NOTE

1. Myers, Thomas, *Anatomy Trains*, 3rd edn, Edinburgh, Churchill Livingstone (2014), p.83

CHAPTER 4

THE HORSE'S SUPERFICIAL FRONT AND BACK LINES AND RIDING IMPLICATIONS

The horse's Superficial Front and Back Lines (SFL and SBL) take much the same form as ours in the trunk, but differ of course in the legs. The horse's trunk is horizontal, so when we talk of the horse's 'topline' we really mean his SBL. There are, remember, left and right lines, which the veterinary researchers collectively called the dorsal (meaning back) line. We can see that this extends from the plantar cushion of each hind foot, via his flexor tendons to his hocks, and via his hamstring muscles to the points of his buttocks (his 'seat bones'), his croup, and his long back muscles (which lie under the panels of the saddle). Each side then passes under his shoulder-blade, along the top of the vertebrae in his neck, and over his poll to the temporal muscles just above his eyes. Finally, it passes into the fibres of each masseter muscle – the chewing muscle which covers each side of his jaw (see Figs. 4.1, 4.2, 4.3).

As in us, the horse's SBL (dorsal line) lies deeper than his gluteal muscles and passes through sheets of fascia that cover his pelvis. Both species have a ligament running along the top of the spinal processes, which becomes the nuchal ligament in the neck. This is much more developed in the horse than in us, and it connects the highest spinal process of the withers to his skull. From this rope-like (funicular) portion of the ligament, the sheet-like (lamella) portion extends like fingers reaching down to the vertebrae of C2 to C5. (C1 is just behind the poll, and like us the horse has seven neck vertebrae.)

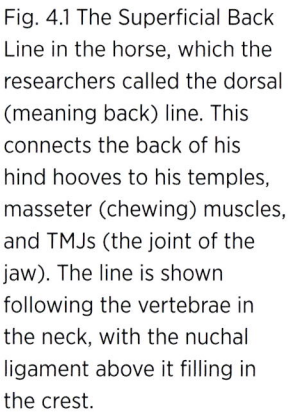

Fig. 4.1 The Superficial Back Line in the horse, which the researchers called the dorsal (meaning back) line. This connects the back of his hind hooves to his temples, masseter (chewing) muscles, and TMJs (the joint of the jaw). The line is shown following the vertebrae in the neck, with the nuchal ligament above it filling in the crest.

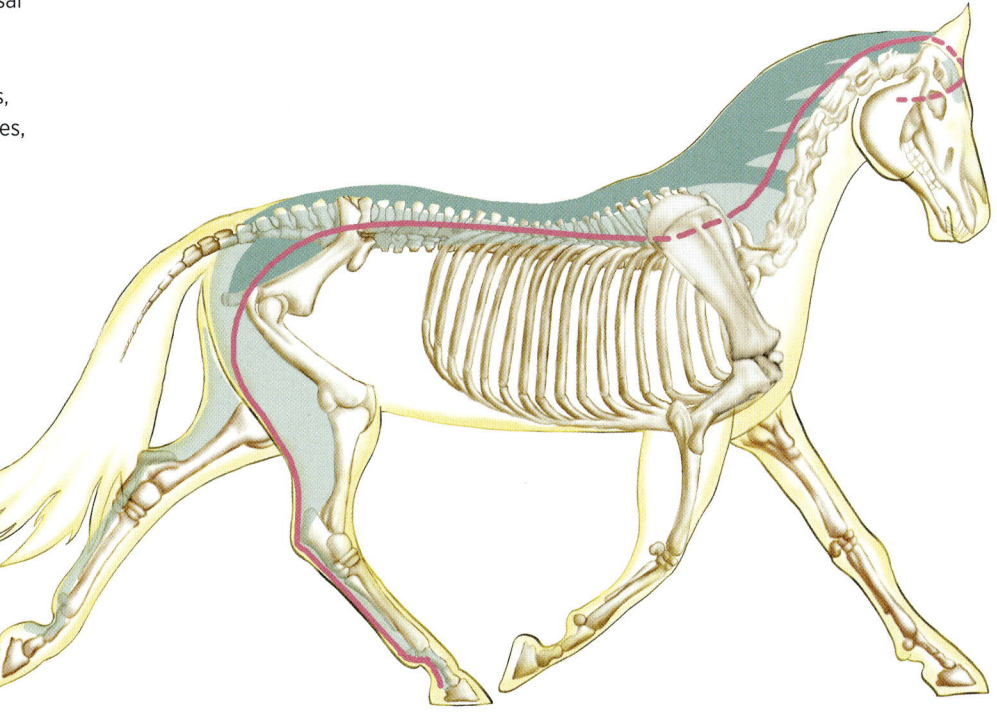

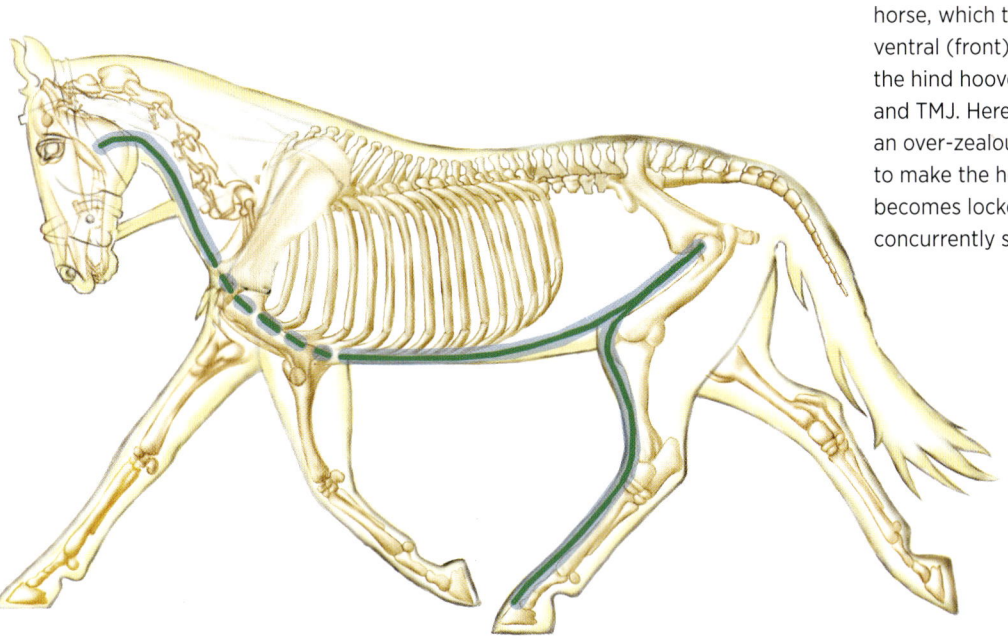

Fig. 4.2 The Superficial Front Line in the horse, which the researchers called the ventral (front) line, connects the front of the hind hooves to the masseter muscle and TMJ. Here it has been 'scrunched' by an over-zealous rider who is determined to make the horse's nose vertical. When it becomes locked short, parts of the SBL are concurrently strained and locked long.

The elasticity of the nuchal ligament enables the horse to reach down to graze, and then lift his head with virtually no effort. In the alert position the ligament is slack,[1] and in movement it stores elastic strain energy that both *limits* the movement of the head and neck in the various gaits, and contributes to the work of moving them. The human SBL includes this ligament, and Myers believes that we should include it in the dorsal line/SBL of the horse, although the veterinary researchers did not do so.

This ligamentous connection from sacrum to poll begins further back still, in the ligament that lies above the vertebrae in the tail. It also extends into the back of the hind leg via one of the hamstring muscles, the semitendinosus, which contains a tendinous band (the clue is in its name!) and joins two vertebrae just above the obvious beginning of the tail to the Achilles tendon in the hock. Thus the entire ligament system above the horse's spine links hock to poll, and it acts like the wires holding up a suspension bridge.

Fig. 4.3 The Superficial Back and Front Lines are in balance when the horse is in the carriage of Fig. 4.1 and also when he is reaching forward, down and out, as here. The bones of the neck then form one curve, as do his crest and Superficial Back Line. There are many possible places of balance in between these options; but many riders inadvertently create the carriage of Fig. 4.2. Some riders deliberately make the Superficial Back Line much longer than the Superficial Front Line.

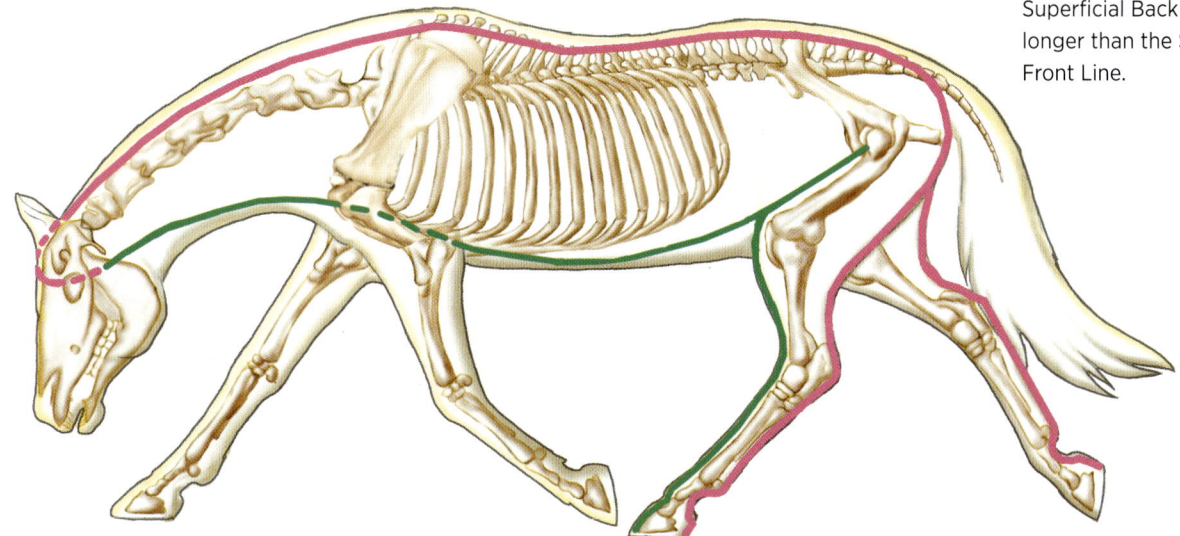

The stanchions of the bridge are the withers and the croup, and when the horse's abdominal muscles contract, the ligament system lifts the spine. The simpler 'bow and string' model proposed in the 1940s compared this mechanism to an archery bow. The lifting itself is passive, since ligaments cannot actively contract like muscles. Realise that if the horse's long back muscles contract his back hollows, and these muscles (like all muscles) cannot *make themselves* lengthen; they can only be lengthened via the contraction of the opposing abdominal muscles. When we consider the horse's core muscles in Chapter 8 we will discover that they, too, have an important role in the lifting of his back.

The entire SBL shortens when that hind leg is flexed. This affects the line further forward, which shortens even more when the back is hollow, and the head up. The SBL on one side lengthens in its entirety when the horse brings his hind leg under him, lifts his back and reaches his neck out of his withers, finding the carriage of 'equipoise'. The dressage term 'over the back' refers to a reach in the horse's long back muscles, and the passage of energy through them. This becomes blocked when the horse has a pattern of holding tension in his loins, under the saddle, in his withers, or in his neck, and over time, this leads to unhealthy fascia.

As riders we are searching for efficient, easy force transmission throughout the whole SBL. The long back muscles under the saddle can allow this whilst also having some isometric contraction. This does not make them shorten – they remain the same length whilst becoming firmer, and giving the rider a more supportive surface to sit on. Skilful riders can feel if one side of the SBL is more compromised than the other, and can influence this. Understanding the line its entirety makes us appreciate that a problem anywhere along it – and also at either end of it – can appear as a constriction at any point along its length, regardless of whether it is caused by injury, bad saddle fit, or issues with shoeing or bitting. We do our horses a disservice by viewing one part of the line in isolation.

The horse's SFL (or ventral line) begins on the front of the hind hooves and passes via his extensor tendons to his hocks, stifle and pelvis. Like ours, it then begins again from his pubic bone, going along the belly through his 'six-pack', between his forelegs and up the muscles on the bottom of his neck to his ears. It passes just beneath (not behind) them into his masseter muscle, where it melds with fibres from the SBL (see Figs. 4.1, 4.2, 4.3 pages 82-83).

Talented dressage horses have a naturally good balance between the SBL and SFL, and can find the equine version of bearing down, where the pressure in their insides increases, and the SFL shortens enough to become a wall that supports their guts and spine. Like human mothers, many broodmares who later return to work can have a long, weak SFL, so their belly sags and their back becomes hollow. Many older horses have the same problem.

Thoroughbreds, Arabs, and other particularly flighty horses also favour the carriage produced by a short SBL and a long SFL, which increases the risk of kissing spines. The startle response in a horse shortens the SBL, mirroring our Landau (extension) response, and rehabilitation programmes for afflicted horses always stress the need for them to work in a way that tones and shortens the SFL, developing tensegrity and 'equipoise' as the SBL lengthens. If you have diagnosed yourself well and taken some of my earlier suggestions to heart, you have already begun the process of developing those skills!

In all quadrupeds, when the SFL shortens and the back arches in flexion, *the face and eyes naturally stay in contact with the outside world*. But this is *not* what happens when stretching the horse's SBL and shortening his SFL in the position known as 'deep and round'. The natural reaction is inhibited even more in the extreme of 'rollkur', which elongates the SBL and shortens the SFL to an extreme that is *not* encouraging muscles to work well in their mid-range, and is not developing tensegrity. Indeed, this does to the horse what we are doing to ourselves as we slouch over our computers! This pattern leads to a very submissive but un-free horse.

Stretching the horse 'forward, down and out' (see Fig. 4.3 page 83) maintains more balance between the length and tone of the SFL and SBL. However, it is more difficult to achieve, and many riders become addicted to the feeling they produce when the horse's back is raised so much that the curve in his crest puts his nose close to his chest.

The phrase 'over the back', which has been in vogue for many years now, encourages riders to think only about the horse's SBL right under the saddle, and not to concern themselves enough with the balance between the SBL and SFL, and also with the deeper muscles of the horse's core. As a result you can expect horses to develop unhelpful patterns such as extra fascial strapping either in front of the withers or higher up the neck, which will, in turn, affect the natural stride of the forelegs.

There are some horses who would, of themselves, adopt this carriage. They are just like the round-backed human, and they have the dubious specialty of being able to 'pop' a rider up off the middle of their back, and thus render her ineffective. It takes skilful riding to not be disorganised in this way, and indeed to rearrange the horse so that his back is more level and his SFL longer, with his neck and poll higher, and his head more out. This is one of the rare occasions when riders have to think in terms of levering themselves *down* onto (you could think 'into') the horse's back rather than working to support their own bodyweight, and to sit lightly.

From this you can see that there is not one perfect bit of advice, no 'one size fits all'. You really have to see your horse's pattern, and discover your own unique advantages as well as the problems that prevent you from finding a feeling of 'oneness' with him. I hope you agree with me that this feeling is worth the work with both 'animals' – you and your horse.

Hyperextension in Horse and Rider

At the opposite end of the spectrum from the round-backed scenario just described, the rider who elongates her own SFL cannot stop the horse from elongating his: thus the hyperextended rider encourages hyperextension in the horse.

Balancing the lines in the rider encourages the horse to come into carriage; however, the reverse is also true, and it is harder for the rider to find the appropriate balance if the horse is out of true. His posture pitches her forward or back; to avoid this she has to learn how to produce extra stabilising tension in one of the lines. This is achieved by finding more robustness in the fixes *already being used* to take her from default posture to neutral spine.

All of our respective SFLs and SBLs are like chains that are only as strong as their weakest link. Our personal weak link determines how our body will react when the horse sticks his nose up in the air and 'pushes back' at us. Think of this like someone putting her hand on his muzzle, pushing his muzzle back towards his poll, his poll back into his neck, his neck back into his withers, and his withers back into his back. His back makes a hollow, and we then fall (according to our weakest link) into that hole which I call the 'mantrap'.[2] Fig. 3.5 pages 55-57 shows all of the possible options.

It is a resilient, responsive rider who can remain in neutral and bring the horse back to neutral as well. There is huge skill in being able to find that extra stabilising tension – especially as the change in the shape of the horse's back always plays right into our pattern of weakness. It requires a version of strength that is totally foreign to most riders, and it is particularly challenging to those who are loosely strung. The last part of the last exercise in Chapter 3 (see page 81) demonstrates beautifully how this strength feels in the SFL. The same resilience is also required in the skills of surfing, snowboarding, skateboarding and skiing, since all of these athletes share the challenge of stabilising their body on top of a slippery or changing surface.

If the horse yanks on the reins, dramatically lengthening both his SBL and SFL, your body will again respond according to your weakest links. Only a well-balanced, resilient rider (with the ability to bear down strongly!) can remain in place without any deformation in her SFL and SBL. She will have a stronger contact on the reins as the horse pulls, but this will not make her arms ache as it is her torso that takes the strain. For the horse, this feels is as if he is pulling against side-reins. Realise that your horse is like your three-year-old toddler – if his annoying games irritate you, he is likely to repeat those tactics. When the horse cannot displace the rider, pull her about, or make her pull back (i.e. when his antics yield little reward), he quits doing them sooner than you might expect.

The rider's task is to balance the tone of the horse's SBL with that of the SFL, and to 'sell' this movement pattern to him. As the horse develops strength, the long back muscles that lie beneath the panels of the saddle become a firmer 'bridge' between his quarters and his forehand. You begin by balancing the lines in yourself, and your next task is to make both his left and right SBLs function more equally, which makes for a straighter horse. If you can organise each SBL as it extends forward in front of you, and also back behind you, you potentially gain access to the entire length of his body from his hind legs to his ears and jaw!

Extending your awareness in this way requires quite some skill. As well as equalising your own SFL and SBL, you have to stack them over the horse's lines *without any side bends or other 'kinks'*. Chapter 5 on the Lateral Lines begins to address these issues of side-to-side balance.

Rising Trot Mechanism – Ballistic Training for Horse and Rider

A good rising trot mechanism is a tremendous tool. Carl Hester uses only rising trot on his horses for the first two years of their training, feeling that they have so much to learn, and do not need the added demand of carrying a rider who is sitting. Rising trot is easier for both horse and rider, and it has the advantage of being ballistic training for both of them. Ballistic exercises, long banished from the curriculum, have now been shown to have value in making your sinews more springy and resilient. The skilled rider uses rising trot to get the tension in the horse's fascial net in synch with her own, so together they get in rhythm and 'play the right note'.

To understand how this dynamic works, imagine two people sitting opposite each other and playing a game of catch with a tennis ball. Each throw involves bouncing the ball once. This is a cooperative game, with its own distinct rhythm, in which neither person is trying to catch the other one out. 'Bounce, catch' mirrors 'sit, rise', and our analogy is essentially describing the exchange of force and energy between horse and rider.

Imagine the human game in full swing – until one person cleverly substitutes a bean-bag for the tennis ball. When this barely bounces, it marks the end of the game! Or perhaps one person suddenly substitutes a 'boingy' ball, and the game immediately speeds up as the ball travels faster.

The skilled rider maintains the equivalent of a tennis ball game, even when the horse would rather throw bean-bags or 'boingy' balls. It is as if the rider says, 'Sorry horse, but whatever you attempt to throw me, I am throwing back a tennis ball', maintaining this resolve and technique until such time as the horse takes a deep sigh and agrees to throw tennis balls!

Whilst some heavy horses have wonderful 'boing', most 'bean-bag' horses are from the heavier breeds. They trot as if their legs were stuck in porridge, since the recoil energy in their tendons and ligaments – and the force transmission along their respective lines of pull – is not enough to enable them to 'ping' off the ground. Iberian horses can be like this too, and any horse with a 'soggy' fascial net will be heavy on the ground and lack the spring of elastic recoil. We could say that the 'bean-bag' horse is too 'down loose'.

Most riders fall into the trap of throwing a bean-bag back to this kind of horse: they land heavily in the saddle and press down into it. This is incredibly instinctive, and is encouraged by phrases like 'sit deep and drive the horse forward', but a heavy landing will inevitably deaden your horse. The game then becomes a vicious circle as the 'bean-bag' tendencies of each partner amplify those of the other. Understandably, most riders soon start to feel like a desperate, disgruntled, and thoroughly jangled bean-bag! The only answer lies in landing lightly and quickly – in a trampoline-like way – thus encouraging the horse to be lighter and quicker on the ground. The rider may also need to kick, and to tap with a whip, but she must not throw bean-bags or she will get bean-bags back.

In contrast, 'uptight' Thoroughbreds throw 'boingy' balls. The rider who gets 'boinged' out of the saddle by the horse's 'boinginess', loses control of the tempo,

(a)

Fig. 4.4 In rising trot the rider is down when one diagonal pair of legs are in mid-stance, and up when the other pair are in mid-stance (a) and (c). She is halfway between in (b) (the ideal moment for a photograph). Notice how the upper body angle changes, so she is leaning slightly forward in (a) and (b), and is vertical in (c). The calf remains still throughout, as the thigh acts as a windscreen wiper. Good biomechanics being shown by teenager Millie Dove, who is learning this schema, and Olympic rider Heather Blitz.

(b)

(c)

and the game speeds up *unless* she can make *a momentary pause on each landing* (without becoming soggy like a bean-bag). This keeps the horse's feet on the ground for a fraction of a second longer. If the rider can also make a pause at the top of the rise (when the other diagonal pair of legs are in mid-stance) this again will act to keep the horse's feet on the ground for longer. This slows the horse's tempo, and changes how his fascial net rebounds from the ground. He has no choice but to throw tennis balls back to the rider (see Fig. 4.4).

This skill can only be built on a really good rising trot mechanism. To understand this, imagine that the rider's thigh moves as if it were a windscreen wiper, rotating around the knee as its pivot point (see Fig. 4.4). Her pelvis moves much more than her shoulders during the rising and sitting; the angle between thigh and torso opens on the rise and closes on the sit. Nothing from the knee down changes in either case, which means that she must not push off from the stirrup and send her foot forward whilst rising. Neither must her foot come back as she lands, making a kick or nudge (to which she might be completely oblivious). The horse will soon become oblivious to it as well, and indeed a rider doing this is training him not to take the leg aids seriously. The rider's SFL and SBL need to stabilise her calves throughout the motion, so that all of the action happens above her knees, whilst her feet remain still and rest lightly in the stirrups.

At the top of the rise, the rider's torso should be vertical; however, it needs to be inclined slightly forward in the sit. Otherwise she would lose control of how she lands, and also struggle to get out of the saddle again. The rider's hips joints open with the rise and close in the sit, but we want the hinge of the mid-back to remain neutral. However, hollow-backed riders

(a)

(b)

(c)

will be very tempted to hollow more as they rise, and if they then land hollow in the saddle, they will be 'pinged' out of it easily. Round-backed riders may become neutral as they rise (although some become hollow), but they then tend to round their back as they sit, and this can make them land too heavily.

This might make you suspect that round-backed riders might slightly deaden a 'boingy' horse, whilst hollow-backed riders – especially the emotionally 'uptight' ones – might do better on lazier horses. Whilst we will all have our favourite type of horse (who reacts well to the tension in our own fascial net), what we really need are skills that broaden our repertoire, helping us to 'read' and adjust any horse we are given to ride, using how we rebound from the saddle to adjust how he rebounds from the ground.

Matching the Forces in the Thrust of Rising Trot

It is a fundamental tenet of biomechanics that the rider must match the forces of the horse's movement, and rising trot is the easiest way to learn this. (In sitting trot, the rider must do this whilst appearing not to move – and this requires more skill and experience.) The rider leaves the saddle as the horse's inside hind leg thrusts off the ground, and when the rider thrusts less than the horse, he has the experience of towing her along.

The vast majority of riders are taught to under-thrust, by 'Keeping the rise small and tidy'. Very few riders over-thrust after the initial stages of learning. When the rider under-thrusts, her centre of gravity is left behind the horse's, *and she becomes the water-skier to a speedy horse's motorboat.* As the horse tows the rider along he throws balls that are faster and 'boingier'. These horses are like type A people (who race around and have heart attacks), whilst the lazy 'bean-bag' horse is like the Type B 'couch potato'. He may happily decide that an 'itsy-bitsy' trot is the best answer to his rider's 'itsy-bitsy' rise. They both agree to play an under-powered game with tiny throws; but you will rarely 'sell' the 'itsy-bitsy' idea to a highly strung 'motorboat' horse!

Suppose the rider needs to slow down the legs of a horse who is speeding off with her. The instinctive answer is to think 'whoa' and to make a smaller 'soggier' rise. But then her centre of gravity is left *even more* behind, and the 'motorboat' speeds off even more! If the rider cannot slow the horse by making pauses, the only remaining tactic is to slow the speed of the windscreen wiper thrust *whilst still getting to the top of the windscreen,* i.e. whilst also matching his thrust and reaching the correct balance point. This requires the rider to rise and sit *as if against a resistance;* perhaps imagining moving through water, or even thick treacle.

This has profound implications. Imagine a rider who is not brave enough to commit to 'going with' a speedy horse, and who leans back whilst thinking 'whoa'. By the laws of physics, *this rider is telling the horse to 'go'*. Similarly, if a skier leans back whilst thinking 'I really don't want to go down this mountain…', this guarantees that the skis will go out from underneath her even faster. Fear makes our worst predictions come true!

Whether on skis or on a horse, it is challenging for many people to commit to *matching the forces of the movement and staying over the correct balance point.* For many riders who aspire to riding dressage, this is an act of courage and not what they thought they signed up for! As you learn how to organise your fascial net, your increasing stability will give you the courage to be 'up for it'. Your horse will thank you, and your riding will become a communion with him, not a battle.

It is all too easy for riders to get caught up in the water-ski/motor boat dynamic, and be carried off with little control. It is equally easy to accidentally deaden a lazy horse's 'trampoline' with a heavy, 'dead' landing. Elite riders take for granted that they can soon make any horse's trot into a 'tennis ball game', and because of the curse of unconscious expertise, they often cannot help their pupils solve the problems of bean-bags and 'boingy' balls.

When riders learn to balance their SFL and SBL, and at the same time to breathe and bear down, they begin the process of becoming stronger and more resilient, able to tension their (and their horse's) fascial net appropriately. This makes them much more able to 'call the tune' with respect to the horse's carriage, the speed of his step, and the spring in his step. They have the same challenges, of course, in sitting trot; but rising trot is ideally suited to skill-building. Lesson one is to build the balance between the SBL and SFL – on both sides and in both of you – and this balance of tension between front and back will change the horse's movement and make his back much easier to sit on.

CHAPTER 4 NOTES

1. Gellman, K.S., and Bertram, J.A.E., 'The equine nuchal ligament 1: structural and material properties' and 'The equine nuchal ligament 2: passive dynamic energy exchange in locomotion', *Veterinary and Comparative Orthopaedics and Traumatology* (2002), 55 15(1):pp.1–6; 7–14

2. Wanless, Mary, *Ride With Your Mind Essentials*, Kenilworth Press (2001)
   Wanless, Mary, *Ride With Your Mind Clinic*, Kenilworth Press (2008)
   Wanless, Mary, *The Naked Truth of Riding*, DVD set, Self-published (2014)

CHAPTER 5

THE LATERAL LINES IN RIDER AND HORSE – THE INTERMEDIATE AND OUTER STABILITY SYSTEMS

The Lateral Lines in the Rider

We will now presuppose that the Superficial Back and Front Lines are already well stabilised, and turn our attention to the two Lateral Lines (LLs), which run down the sides of the body from the ear to the outer arch of the foot. They form what we could call the 'outer stability system' of the body, and also give rise to the 'intermediate stability system'.

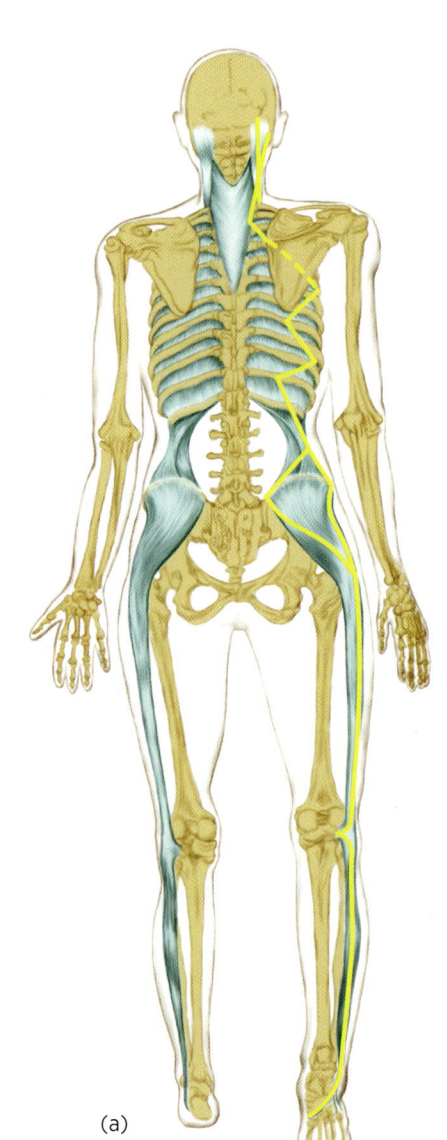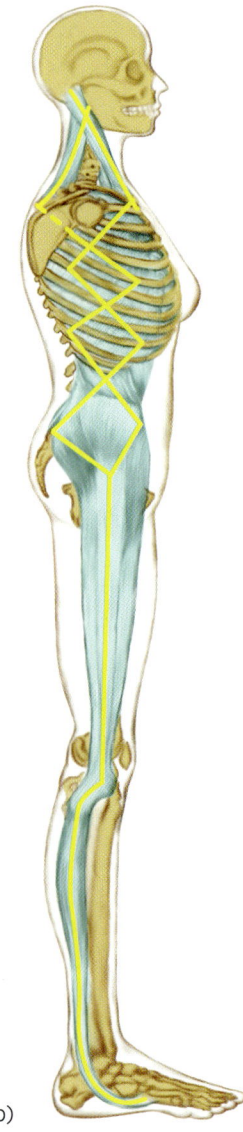

Fig. 5.1 The Lateral Liness from the back (a) and side (b), showing the 'plimsoll lace' 'Xs' lacing the sides to the Superficial Front Line and Superficial Back Line at the front and back.

As with the Superficial Back and Front Lines, I will describe the LLs from the bottom to the top (see Fig. 5.1). However, in practice, tension is passed along the lines in either direction. With riders in particular, it may be directed both towards and away from the waist, as the fascia on one LL pulls in on itself and locks short, whilst the other LL becomes locked long.

Each LL takes root from two tendons, one of which inserts into the joint just up from the big toe (at the base of the first metatarsal) and the other into the equivalent joint of the little toe. Both tendons pass around the outside of the mid-foot, supporting the lateral arch, and hooking behind the obvious bony knobble of the ankle (the malleolus). The line continues through the muscles of the outer calf which blend into the iliotibial tract, a very strong sheet of fascial fabric that you can feel just above the outside of your knee. This continues up the outside of each thigh as a defined band that is most visible on skilled, fit male riders, but also there – and strong – on women.

The band widens out to cup the bony knobble of the thigh bone (the greater trochanter of the femur). This is what you land on if you fall sideways skiing, and it is easy to feel (see Fig. 5.2). From here the line becomes a 'Y' shape, whose branches attach to the front and back of your iliac crest, the bony ridge that your hands rest on when you put them 'on your hips'.

From here up, the LL continues as a series of 'Xs' similar to plimsoll laces, in essence lacing together the front and back of the torso. There is a large 'X' criss-crossing your sides in the space between your ribs and hips, through the external and internal obliques. Between each rib lie two layers of muscle (the intercostals) that again criss-cross, forming a series of tiny 'Xs' like basket weave. These tie

Fig. 5.2 Find the greater trochanter of the femur by putting your hands on your panty line and leaning to one side. It should become clear on the side you are leaning towards.

Fig. 5.3 When seen from the back, most riders have one bigger, rounder-looking buttock, which tends to fall off the side of the saddle – this is the 'girlish' side, seen here on the left.

all your ribs together all the way up your sides, but soon become covered by the large shoulder muscles. But the LL re-emerges in your neck with the last 'X', two broad muscles that work to keep the head balanced and steady as the body moves underneath – an active process in both running and riding.

In a fish, the LLs provide the primary motive force for locomotion, but the LLs in human beings form an adjustable stability system on the outside of your body, activated on the supporting side as your foot makes contact with the ground in each step. Adjustability and a ready ability to tone these lines are necessary to the undulating skier, the dodging footballer, and the responsive but stable rider.

It would be so helpful if we could all begin our riding career with two equally toned LLs! Unfortunately, we distort them whenever we carry a heavy shoulder bag, soon reaching the point where the bag only sits comfortably on one shoulder. Carrying a young child consistently on one hip creates even more profound distortions. When you look at a static rider from the back – even a young one with little experience of bags and children – you will almost always see one side of the pelvis having a more 'girlish' cast, whilst the other side can look more 'boyish' (no sexism implied). The 'girlish' side, which has the larger, rounder, more feminine-looking buttock, is the side of the longer LL (see Fig. 5.3).

The Lateral Lines in the Horse

The LLs in the horse (see Fig. 5.4) look very similar to those in the human, lacing his sides and similarly helping in side-to-side stability. They pass from just behind the

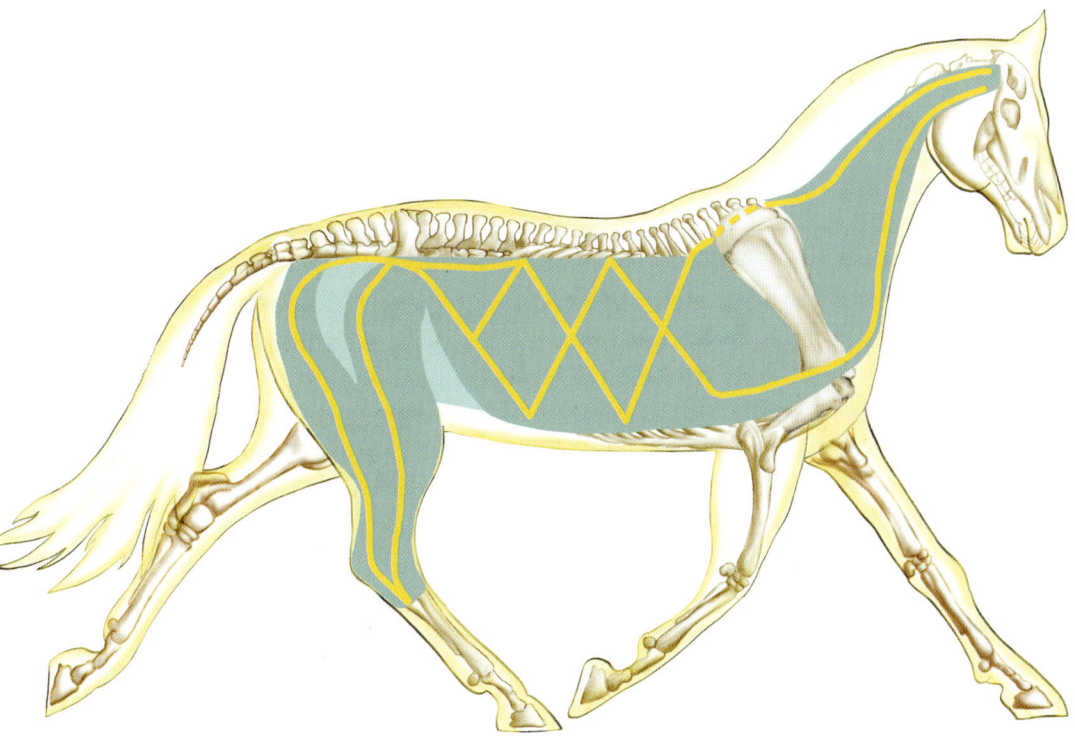

Fig. 5.4 The Lateral Lines in the horse mirror those in the human, making 'Xs' along his sides. If his back is hollow, the upper set of 'Vs' act to shorten one side and elongate the other. When he is lifting his back and reaching into the reins, the bottom set of 'Vs' shorten one side and elongate the other. You can mimic this in your own body by hollowing or rounding your back as you side bend.

ear, through the neck to the small muscles between the many ribs, and also to the lateral abdominal obliques which form the 'X' behind the ribs. In the hindquarters, the LLs turn 90 degrees (not true for us with our extended hips), and go down to the outside of the hocks (the equivalent of our heels).

The veterinary researchers think of the LLs as 'Vs' rather than 'Xs', and call the line of the upper inverted 'Vs' the Deep Lateral Line. It passes under the shoulder-blade and though deeper muscles of the neck, creating a side bend in the horse when his head is up and his back is hollow.

The lower 'Vs' they consider the more Superficial Lateral Line (SLL), which passes not only through the small muscles between the ribs, but also through the large and important cutis lateralis. This is the muscle your horse uses to shake of flies in a manner no longer available to humans! The cutis muscle is literally in the skin, but at 1.2 inches thick it is large and strong enough to participate in the horse's side-to-side balance. Furthermore, it links the LL to the outside of the hock, suggesting the mechanism through which our lower leg can influence the horse's hind leg.

This superficial branch of the LL passes over the lower end of the shoulder-blade, again through the cutis muscle, and then through more superficial muscles of the neck to meld with the upper line just behind the horse's ear. As those lower 'V's come together, the LL side-bends the horse *and also flexes his back and neck,* influencing his carriage.

Imbalance between the LLs on each side leads to an uneven rein contact, and difficulties in turning. A locked short LL on one side will make it impossible for the horse to bend to the opposite side. Many riders would think of this as the primary issue, but I think of it as the problem *beyond the initial problem*, which lies in his tendency to 'hinge' at the withers.

When the horse falls out, his withers acts like the hinge of an articulated lorry (an eighteen-wheeler), and he 'jack-knifes' (see Fig. 5.5(c) page 97). His torso then follows his withers, and you probably know from experience that any attempt to steer by pulling on the inside rein is doomed to fail! The horse's outside LL has become overstretched, as has his outside Superficial Back Line: they are *strung out*, and locked long, reminiscent of some kid pulling her chewing gum into a long thin string (Sorry!).

When the horse's locked short side is on the outside, he will tend to fall in – and whilst this too presents a challenge, the rider's instinctive response is less damaging than it is on the rein where he falls out.

Fig. 5.5(a), The most bendy part of the horse is his neck, and the next most bendy part is from the eighth to the sixteenth thoracic vertebrae. This reaching around involves the Lateral Lines and also the Spiral Lines (see Chapter 8), which rotate the torso and neck. In us too, rotations normally accompany a side bend.

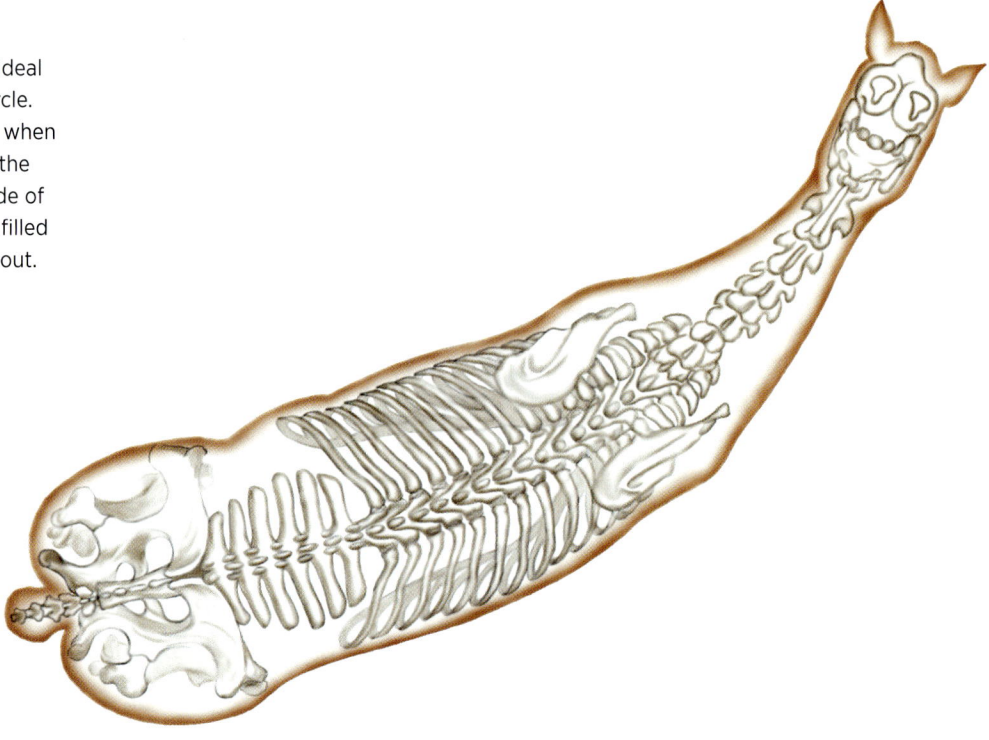

Fig. 5.5(b), Shows the ideal bend on a 20 metre circle. This becomes possible when the rider can maintain the boundary on the outside of the horse, making him filled out rather than strung out.

Fig. 5.5(c) The jack-knife, where the horse hinges just in front of the withers rather than adopting the bend that results from smaller lateral displacements in many more joints. This very often happens when the rider pulls on the inside rein: the horse's nose is pulled to that side, whilst his withers and torso go the other way!

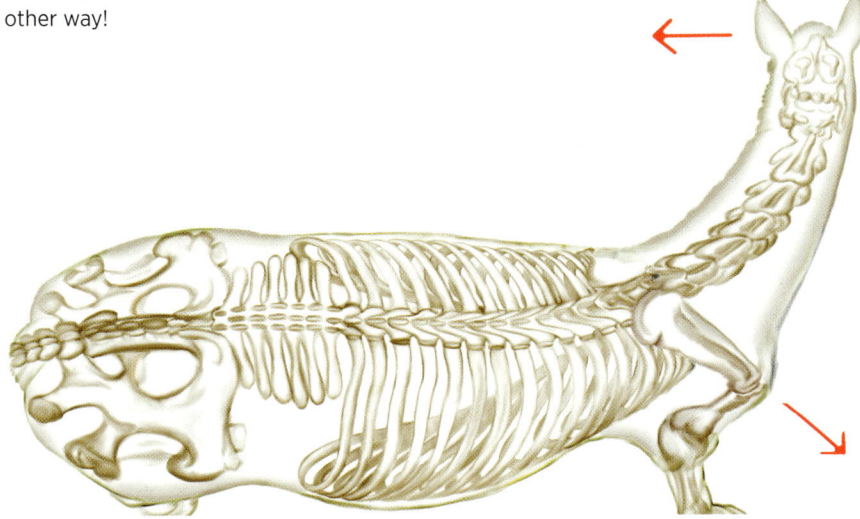

The Ribcage and Hind Legs

The difference between the two LLs is often obvious when a young horse is first worked on the lunge, and most trainers describe him as having a 'stiff side' and a 'soft side'. The rein contact may be rather strong on the stiff side, and close to non-existent on the soft or hollow side. The horse neither weight-bears equally, nor pushes off equally, with each hind leg. This imbalance usually extends to unequal Functional Lines and Spiral Lines (see Chapters 6 and 8), and keeps his ribcage held to one side – often seen in the lack of an equal swing from side to side.

Many equestrian authors suppose that the hind leg on the side that the ribcage bulges to is restricted in its movement, but bears more weight. It is often called 'the carrying leg'. The other hind leg can more easily come under the body, but it bears less weight. Some authors call it 'the pushing leg'. This begs two questions. The first is, 'How much are the hind legs influenced by the position of the ribcage, and how much are they (and also the horse's hindquarters) asymmetrical in their own right?' This mirrors the proverbial question about the chicken and the egg! However, in my experience, the rider's greatest influence on this kind of imbalance arises from learning *how to reposition the horse's ribcage*, not from focusing on the hindquarters or hind legs themselves.

Secondly, we have to ask, 'Is the pushing leg pushing weight towards the shoulder on the same side, or towards the opposite shoulder (which is the side of the bulging ribcage)?' The two Superficial Back Lines transmit force parallel to the horse's spine, as would LLs that are equally toned. However, the helical lines of pull (the Functional and Spiral Lines) go diagonally across the spine, suggesting that the net overall pattern can vary from horse to horse and also between different gaits and movements.

Riders have long been told that when the horse is working well on a circle, and also in shoulder-in, his inside hind leg carries more weight. We do not actually know that this is true and Hilary Clayton, Professor Emerita at Michigan State University, has shown in her laboratory that, when a horse is working on the lunge, *the outside fore and hind limbs actually carry more weight* than the inside ones.[1]

Traditionally, we are taught that our riding interventions should address the horse's hind legs to develop and equalise their pushing and carrying power. But if we think of the horse's spine like a train with thirty-two carriages hinged together, that train is not on rails. (Between the poll and the dock there are thirty-two joints between vertebrae, allowing varying degrees of lateral bending. As becomes clear from Fig. 5.5(a) page 96, tiny amounts add up.) So when pushed along from the back the train can become extremely wiggly, especially if the horse's guy-ropes are loosely strung! By analogy, you can push on the end of a broom handle, but you cannot push on a rope.

Thus it serves the horse well to pull himself along from the front, allowing all of his 'carriages' to follow along passively.[2] This suggests that straightness should have a place nearer to the beginning of the dressage 'scales of training', since without it, the 'train' of the horse's body so easily 'derails'.

Dr Andrew McLean, one of the founding members of the International Society for Equitation Science, is a leading proponent of the idea that the forelegs are *at least as important as the hind legs*, and should be addressed in their own right, instead of being considered as passive recipients of the weight directed onto them by the hind legs.[3] In fact, the forelegs tell the hind legs what to do, via the 'central pattern generator'. This is a neural oscillator – a cluster of nerves lying within the spinal cord, which acts like a light switch with multiple settings. It coordinates the rhythmic motions of limb movements. He argues passionately for a revision of the training scale.[4]

Bendy Body Parts

The horse's neck is his most bendy body part, with most of the movement happening between the poll and the atlas (see Fig. 5.5(a) page 96). This joint allows a flexion to the inside – if the horse had a unicorn horn, it could point up to 40 degrees each way just on this first joint, which at its most upward point creates a dimple a few inches behind his inside ear. His next most bendy joint is just in front of the withers, between the two lowest neck vertebra (C6 and C7). This, along with its neighbouring joints, creates the hinge 'at the withers', which allows the horse to jack-knife (see Fig. 5.5(c) previous page).

The tiny amounts of lateral displacement between the vertebrae of the ribcage and loins have allowed us to invent the various lateral movements, which we will consider the Chapter 8 on the Spiral Lines. However, they also allow for some creative misinterpretations on the horse's part – especially if his lateral 'guy-ropes' are rigidly stuck at different lengths, or so loosely strung that steering him is like trying to line up a pile of noodles!

Many books have diagrams similar to Fig. 5.5(b) page 96, which shows the ideal bend on a circle; but they do not show the jack-knife of Fig. 5.5(c) page 97. There is a huge difference between the two, and it does not serve us to pretend that 5.5(c) never happens! My experience is that unless the rider is unusually skilled, she needs to learn how to *equalise* the horse's two LLs (and contain that jack-knife) before she can learn about 'bend'.

My approach in this chapter is designed to help you to steer the horse's withers on the track of your choosing, by positioning the withers, neck, head and ribcage so that his two LLs become more even. This, in turn, helps to make the asymmetrical ribcage more even and, by proxy, it affects how the hind legs step, since they are not then inhibited by the imbalance. This method is unusual, since it is based on your need to even out *your* two LLs as you discover how to even out your horse's – but it is a very practical way to help you to steer and, unlike most discussions of 'bend', it does not presuppose skills that you do not have. (We will arrive at this point in Chapter 8 on the Spiral Lines; resist the temptation to jump head of yourself!)

Rider-Horse Interaction

Our best intentions as riders do not always manifest in reality. Steering, in particular, brings out those discrepancies, and the 'default' created by our unequal LLs influences the horse in ways that are often blatantly obvious, but sometimes subtle enough to pass underneath the radar of our awareness.

All too often, the combination of our default and our horse's default can be a problem squared, not a problem doubled (and sometimes 1+1= 11!). This means that groundwork is often a helpful way to reduce the effects that unequal LLs have on the horse's carriage and movement, and even on his temperament, since the most asymmetrical horses are often quirky, as well as predisposed to lameness. Also, when we are on the ground, the effect of our asymmetry is minimal compared to when we are riding. But since increasing our ridden skills requires us to open the Pandora's box of the LLs, here goes …

Side Bends and Hot-air Balloons
If we think of the 'X's of the LLs lacing the outside of the rider's torso like shoelaces, we can then think of eyelet lines defining where the LLs meet the outer edges of both the Superficial Front Line and Superficial Back Line. These boundaries form four strong fascial straps which are approximately vertical and more or less follow the lines of a man's braces (suspenders), front and back. These guy-ropes comprise the body's 'intermediate stability system'. (We will meet the 'inner stability system' in the Deep Front Line of the body's core, in Chapter 9.)

We could compare these straps to the four ropes of a hot-air balloon, which attach the balloon itself (the ribcage and its contents) to the basket (the bowl of the pelvis). As soon as the 'ropes' of the intermediate stability system do not have equal tension, the rider will 'collapse a hip', and this shows most obviously as a difference in length between her two sides, i.e. in the outer stability system of the LLs. Whilst riding on a circle where the locked short fascia is on the inside, the rider's torso will make a 'C' curve to the inside, with creases between the ribs and hips. On the other rein, with the *locked long* fascia on the inside, the rider will usually stay much closer to vertical and steer more effectively.

Do the following exercise, either on a firm chair or gym ball.

1. Cross your hands on your upper chest, and make a side bend with your torso. Notice which side you choose, and how easily that side falls into creases. This is the side that is locked short.
2. Come back to vertical, and then side bend the other way. Almost everyone finds that this is less familiar, and that the torso does not so easily make creases – the fascia is locked long and the muscles are locked into eccentric loading.
3. It is rare that someone does not clearly sense a preferred direction for a side bend. Those who do not, usually have a very square torso, which may look from the back like a square table top. If you are long and willowy you will side bend more easily.
4. If one seat bone becomes heavier and one becomes lighter as you do this exercise, do not assume that you know how this will happen whilst riding – when the surface you are sitting on is saddle-shaped, the weighting of the seat bones may be different (this is explained in page 169).
5. Side bend to the easier side, and add a rotation, twisting so that you advance the shoulder of the longer side. This, too, may have a familiar feel. (It involves the Spiral Lines of Chapter 8, but is worth doing here so that you understand this common addition to the 'C' curve.)
6. Come back to just the 'C' curve, and then rotate that same shoulder back, so you are looking up. I expect that this feels really weird – it is not what spines tend do naturally, given the shapes of the vertebrae!
7. Repeat both of these movements to the opposite side for an even bigger experience of being 'not you'! You might feel some fascial pulls that are very unfamiliar. Challenge the edges of your movement somewhat, but do not risk a pulled muscle.

Nothing in nature is symmetrical and, on a horse, virtually no one's seat bones will naturally carry equal weight. Neither will they be positioned just to each side of the horse's spine: very often one will drop away from his spine, and a small percentage of riders will find that one may even cross it. The rider's spine then curves away from the 'girlish' side of her pelvis (see Fig. 5.3 page 94), and the inner thighs are pulled out of the ideal symmetrical 'V' shape that should encase the horse's mid-line. The rider's knees will not stay hooked onto a horizontal bar (see page 75) – the knee on the inside of the 'C' curve will come up, and/or the other one will go down.

Whenever the rider's torso deforms there are huge knock-on effects throughout the fascial net. Many riders I know have found that bodywork is immensely helpful in changing their pattern – indeed it can reduce the need for years of compensating for your problem, years of trying, and years of not truly succeeding.

Whose Problem Is It?

When riders suffer the associated difficulties of steering their horse, they may kid themselves that they are experiencing issues that are their horse's alone. Of course, horses bring the issues of their disorganised lines of pull to the partnership – but so do their riders. A horse turning by himself will tend to lean in more than we riders find comfortable, and virtually all us have endured the 'wall of death'! However, horses rarely fall *out* when turning on their own, and the rider's contribution to their shared problems tends to amplify this tendency.

On any turn or circle, the centrifugal force tends to throw them both to the outside, mirroring the way that the spin cycle of your washing machine sends clothes to the outside of the drum. When the inside of the rider's body is locked long she can largely withstand this force, and a small percentage of riders will actually lean to the outside. But when the *outside* of the body is locked long (and the inside locked short) the torso is easily deformed (see Fig. 5.3 page 94).

As the rider's outside LL becomes longer, so her inside LL becomes shorter, forming creases between armpit and pelvis, and bringing her chin to the inside of the horse's mid-line. As this happens (as in the exercise above), she is also likely to rotate to the inside, which will face her shoulders and chest that way. This puts her inside hand further back than her outside hand and, before long, she is pulling on the inside rein as the horse jack-knifes and falls out on the circle.

The differences that we have seen in the Superficial Front and Back Lines within the rider's legs (see Chapter 3) also come into play, and on the side where the rider tends to rotate in, the inside heel will usually go forward and down. On the side the rider is rotating away from, her outside leg will usually go too far back and on tiptoe. The sad truth is that, even without the addition of the centrifugal force, the inequality in the rider's various lines of pull creates a force within her body that acts *from the side that is locked short towards the side that is locked long*.

The horse will also be most affected by the centrifugal force when his longer side is on the outside. He becomes *strung out* on that side, and falls out of the circle, jack-knifing instead of turning (see Fig. 5.5(c) page 97). But realise that the rider can only fill out that side if she can *keep it contained within a defined boundary* (see Fig. 5.5(b) page 96). Books and teachers who prescribe 'inside leg to outside rein' are presupposing that your outside rein can easily create and maintain a boundary for the horse's outside LL. The unfortunate truth is that *this will not work* unless your seat bones and thighs are correctly positioned. When they are not, you will *inevitably give that outside hand forward, however much you are told not to!* It is essential to address this piece of the puzzle, lost to our culture through expertise-induced amnesia.

If your asymmetry is running rampant – and especially if you and your horse are both locked short on the same side – steering to that direction will be an amplified drama of falling out on a turn. Riding on the other rein may well present the opposite drama, whereby the horse falls and leans in with his forehand, whilst his quarters swing out of the circle.

With luck, you and your horse will not compound each other's asymmetry quite so dramatically, but you will probably be shocked to discover that I have often known horses to *reverse their stiff and soft sides* within a few minutes of a new rider getting on. This means that, if the first rider can circle easily to the left but struggles to the right, the second rider's influence reverses this pattern so that this rider struggles to steer left. The horse has mirrored that new rider's pattern!

This suggests that the rider's asymmetry can be the dominant factor in how steering pans out. Whilst I have seen and experienced this many times, it will doubtless still surprise some readers to realise the extent to which balance (or imbalance and subsequent compensation) along the LLs of the rider becomes echoed in the horse. In time, the unwitting side-to-side forces of the rider mould him into having issues that most would assume were originally his.

When the horse's pattern is indeed dominant, I believe that he has significant issues with his LLs that make him particularly unmalleable. But whatever the reasons for a significant asymmetry in a horse's LLs, the process of him maintaining that pattern may have different effects on two riders whose asymmetries give them different 'weak links'. However, *nota bene,* the rider's asymmetry is *a much bigger factor than most people want to acknowledge.* As a result, working well it with yields huge improvements to both rider and horse.

Know Your Enemy!

Use the following list of questions to help you identify your asymmetry. Of necessity, they begin with an assessment of how clearly you feel your seat bones.

1. Can you feel your seat bones? Do you have equal weight on each one? If one or both are unclear, put your fingers under the panel of the saddle or under the saddle cloth as in Fig. 5.6 and pull upwards. As you do this, be sure to keep your chin and zipper over the horse's mane. Your seat bone should poke down through your flesh and become clearer.
2. You may need to *let go* in the muscles under your backside to keep this clarity – you could unwittingly be 'popping yourself up'. Slowly contract them and let them go a number of times.
3. Check also that you are not pushing into your stirrups, as this too sends your seat bones up (as in the exercise on page 74).
4. The bones should feel clear on your hands,

Fig. 5.6 Finding the 'missing' seat bone by pulling up on a saddle cloth or the saddle flap. Be sure to keep your torso vertical and facing forward.

but should not poke down into them. If they do, you need to firm up the muscles around them a little so that they too become part of your sitting surface. The underneath of each rider's torso-box is different, and we are all subject to the 'Goldilocks principle'. To find 'just right' some riders need to let go in the muscles around their seat bones, whilst others need to firm them up. Be willing to experiment, and then practise 'just right' whilst driving your car!

When you are clear on this, perform the following checks in walk and trot on both reins on a 20m circle, ideally with a friend who can ask you these questions whilst aiming a camera.

1. Can you keep 50/50 weight on each seat bone whilst in motion? If one seat bone frequently disappears, where does it go? Can you discover how to bring it back?
2. Is one seat bone placed closer to the horse's spine, and one further away from it? Is one further forward, and one further back?
3. Are your chin and mid-line stacked up over the horse's mane, or are they off to one side?
4. Are there creases on one side of your torso? Does your spine make a 'C' curve, or an 'S' curve? The latter is less likely but possible, and is discussed in Chapter 8 on the Spiral Lines.
5. If you had three skewers sticking out of the front and back of your mid-line, as shown in Fig. 5.7, would they all be horizontal? This double-checks that you are vertical and neutral. Would they point straight on over the horse's mane i.e. on the tangent of a circle, or would they point to the inside or the outside of your circle?
6. Is the inside of one thigh lying more snuggly against the saddle than the other?
7. Are your knees hooked on their imaginary bar, with the bar level?
8. Do you have equal weight in each stirrup? Are your stirrup leathers the same length? You need to take them off the saddle and measure them, you cannot trust your felt-sense of this!
9. Do you tend to pull on one rein? Is one rein always longer than the other? Does one wrist or arm 'droop', keeping that rein loopy? Does one of your hands tend to stay closer to, or tend to cross, the horse's neck? Does one hand stay further away from his neck?
10. When you ride a 20m circle, in which direction do you have more difficulty turning your horse? Take your time as you become clear about how your pattern varies when riding in each direction.

Your answers could be very different on the two reins. They form the pattern that describes your asymmetry. Wherever you go, and whatever you do, your aim is to keep 50/50 weight on each seat

Fig. 5.7 If three skewers on your mid-line stuck out front and back, would they be parallel and all pointing straight ahead? When riding a circle they should point along the tangent.

bone, with your knees level and your mid-line over the hose's mane, with no creases in your torso and your skewers pointing straight on.

The problem is that, in your attempts to do this, willpower and obedience will inevitably fail! This is not because you are stupid; it is because once your asymmetry is coupled with your horse's asymmetry and also the centrifugal force, willpower and obedience are no match for the distortions in your lines of pull.

Willpower and obedience will fail too in your attempts to make your hands a pair, as they are only symptoms of problems that lie in unequal fascial pulls within your torso. When you can successfully address the root cause, the symptoms take care of themselves (with help from Chapter 7, which discusses the Arm Lines).

Narrowness in the Thighs and Seat Bones

Before we dive into more depth about the LLs, we need take into account the fact that most riders (and especially most women) naturally sit in a way that makes them *too wide* for their horses. If one thigh and seat bone sit snugly against the saddle, the thigh and seat bone in the other side will tend to fall away from it (see Fig. 5.8(a)). The rider is lacking what I call 'narrowness', and this leads to all kinds of steering problems, which are too often blamed on the horse.

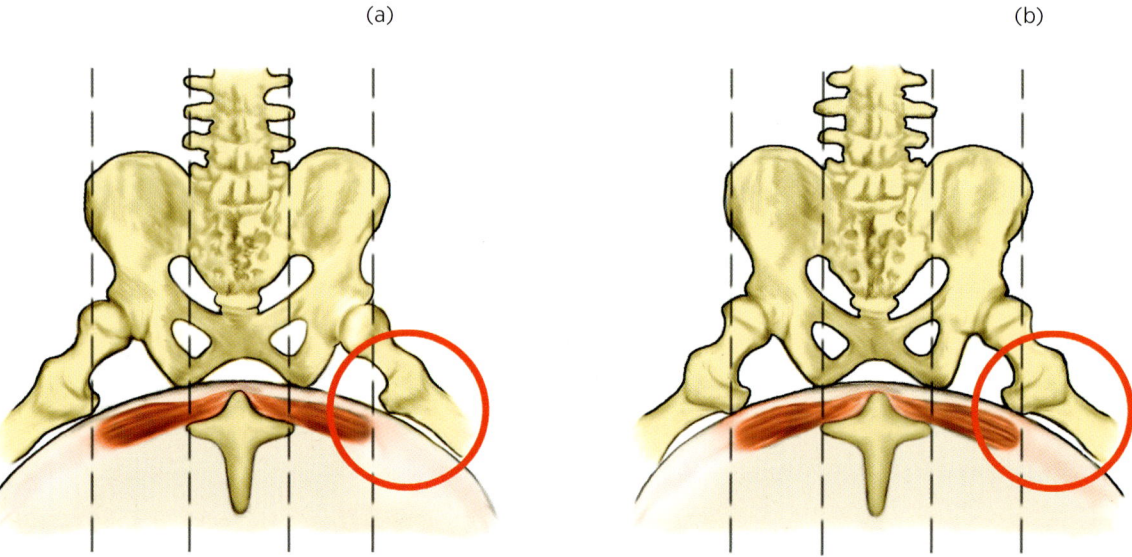

Fig. 5.8 In (a) the rider has one seat bone falling away from the horse's mid-line, and this makes her lesser trochanter (which lies beneath the greater trochanter) fall off the outer edge of the horse's long back muscle. The other side of the pelvis is well placed. (b) Shows an ideal that most riders have to consciously learn to do.

Most riders have experienced the kind of intensive training session that makes such a huge improvement to the 'bad rein' that it suddenly becomes the 'good rein'. The rider will usually experience a few days of euphoria before it dawns on her that the formerly 'good rein' has now become the 'bad rein'! Rarely does a rider realise that the originally loose side has snuggled in against the horse whilst the originally snug side became looser. In fact, some riders make a career out of 'ping-ponging' around on the horse's back, with one side always snuggled in and the other falling away from the mid-line. Others remain stuck in one position for ever; but however frequently you do or do not change sides, the only effective and lasting answer is to *become narrower*.

To steer well, the rider needs to be able to arrange her backside and thighs symmetrically over the horse's long back muscles as in Fig. 5.8(b). This makes intuitive sense, as it affords a specific and symmetrical relationship to those muscles. The rider's backside, thighs, and both sides of the body act like both pieces of bread in a horse sandwich. They limit the 'wriggle room' for his withers, helping to limit the deviations through which he falls in or out. The shape of the male pelvis makes this easier for men, as their seat bones and hip joints are closer together than they are in most women.

This starting point is so valuable that we are going digress briefly to consider the muscles of the inside thighs, which actually belong to the inner stability system of the Deep Front Line (see Chapter 9). Obviously, all the stability systems – the outer LL, the intermediate system, and this inner system – have to work in concert in the end – the separation is for analysis only.

1. Begin from a place where your inside thighs are rotated in, as in Fig. 3.6(b) page 59.
2. Notice which inside thigh lies less snugly against the saddle and, in halt or walk, place your opposite fist on the skirt of the saddle which covers the stirrup bar. Push your knuckles against the saddle, and make a counter-pressure with your opposite thigh (as if you could press your thigh and fist against each other, even though the saddle is in the way – see Fig. 5.9). Notice your thigh firm up, and then maintain this as you take your fist away.

Fig. 5.9 To snug one thigh against the saddle, press the knuckles of the opposite hand against the skirt of the saddle, and make a counter pressure with the opposite thigh.

3. Resolve to keep noticing each time your thigh comes off the saddle, attempting to reduce the time-lag between when it actually happens, and when you notice it has happened! If the correction is not easy, repeat the exercise many times until 'think it' is enough to put your thigh back into place.

There are two other variations on the theme which sometimes work better:

1. Using the hand on the same side as the weaker thigh, place two fingers inside that side of the pommel. Pull outwards with your fingers as you snug your thigh inwards (see Fig. 5.10).
2. Alternatively, put that same fist *across* the pommel and, with the bottom of your fist, press on the skirt as you make a counter-pressure with your thigh. This pressure often affects your seat bone as well as your thigh on the side that your hand originates from, bringing both of them closer to the mid-line (see Fig. 5.11).

Fig. 5.10 Use two fingers on the same side as the weaker thigh to pull directly to the side and snug that thigh into the saddle. This especially targets the tendon at the corner of the pubic bone.

Fig. 5.11 Put the same fist as the weaker thigh across the pommel, and make pressure from your curled little finger. As well as snugging in your thigh on the same side as your arm, this should bring the same seat bone closer to the mid-line.

With one firmer thigh and a soggier one, you can transmit force only in one direction. These resistances all add tone to your less-toned thigh, making your thighs a more symmetrical 'V', and making it possible to transmit force as you choose.

Overcoming Your Instincts

Almost all of us begin our riding lives by attempting to turn in the most instinctive way, which is to pull on the inside rein. Unfortunately, this is about as effective as pulling on the inside handlebar of a bicycle! Whilst cyclists make this mistake only once, riders make it repeatedly because the physics of their situation dictates that when the outside of their body has gone 'walkabout', they have no other choice!

In a profound jack-knife their problem becomes very obvious. However, when the jack-knife is slight rather than acute, less savvy riders are often thrilled about their horse's 'bend'. The idea of being able to see his inside eye makes them blind to the fact that his forelegs are stepping more to the outside than the inside.

Elite riders think much less than average riders about steering the horse's *nose* with the reins and much more about steering his *withers* between their thighs. Imagine the horse's nose as the wheel of a wheelbarrow, and the reins as the handles of the wheelbarrow. The concept is to keep pushing your hands forward, thinking of the reins like the wheelbarrow handles as the nose (wheel) goes along in front of you.

The rider who does this is keeping her horse's LLs even in length and tone, and eradicating the 'wriggle room' for his withers. The result is a horse who stays *straight on the circle,* and who turns *like a bus* instead of like an articulated lorry. The front end turns around the back end, and with no hinge at the withers, he turns as if being steered around a many-sided coin. We will unpack the 'how to' of this very soon.

The common instruction to 'relax your thighs and take your knees off the saddle' makes 'wheelbarrow steering' impossible. This instruction has evolved out of fears that riders will grip; but photographs of elite riders always show the inside of each thigh on the saddle, which suggests that this might be helpful (see Fig. 3.14 page 71).

It works well to imagine your torso and thighs steering the horse's withers and forelegs along an imaginary line that has been painted on the surface of the riding arena. This simple idea can help you resist the temptation to steer the horse's nose, which ensures that you will lose control of his withers. Once your LLs and your intermediate stability system play their part, this can evolve into really precise steering that controls the lengths of the horse's LLs.

The 'Boards' Exercise for the Intermediate Stability System

This next exercise is a huge help in becoming narrow enough to minimise the 'wriggle room' that you unintentionally allow your horse. To add more stabilising strength to your torso, we are going to imagine that on each side, the eyelet line on your front (where the LL joins the Superficial Front Line) could be joined to the eyelet line on the same side of the back (where the LL joins the Superficial Back Line). These connections would form two planes that are parallel to the horse's

(a) (b)

Fig. 5.12 The lines of the boards are as shown on the surface of the body (a) on the front, (b) on the back.

spine. I think of them like two 'boards' joining your front to your back (see Figs. 5.12(a), (b), Fig. 5.13). When these boards can be stacked up over the inner edges of the horse's long back muscles (i.e. the outer edges of the gullet of the saddle) we create a myofascial-chain-to-myofascial-chain connection with the horse that is immensely powerful in organising the dance between you.

The eyelet lines follow the line of a man's braces (or suspenders) so they pass from a woman's shoulder strap, over her bosom, and down the edge of the abdominal muscle, continuing all the way down to the corner of the pubic bone (see Fig. 5.12(a)). On the back they again follow a man's braces/suspenders and from a woman's shoulder strap, they pass down the outer edge of the long back muscle, past the waist, and continue down the line of the sacroiliac joint on the back of the pelvis (see Fig. 5.12(b)). The line continues into the underneath of the torso-box, to the seat bone and the corner of the pubic bone. The boards which join the back and front of each suspender line divide the rider into a right third and a left third, with a core third in the middle.

You will benefit enormously from doing the following exercise off-horse before you attempt the same process mounted. This will give you another coordination to practise as you drive your car, etc., and it will show you how to amplify the stabilising effects of the intermediate stability system. Prepare for some surprises! Achieving this is less difficult than it initially seems.

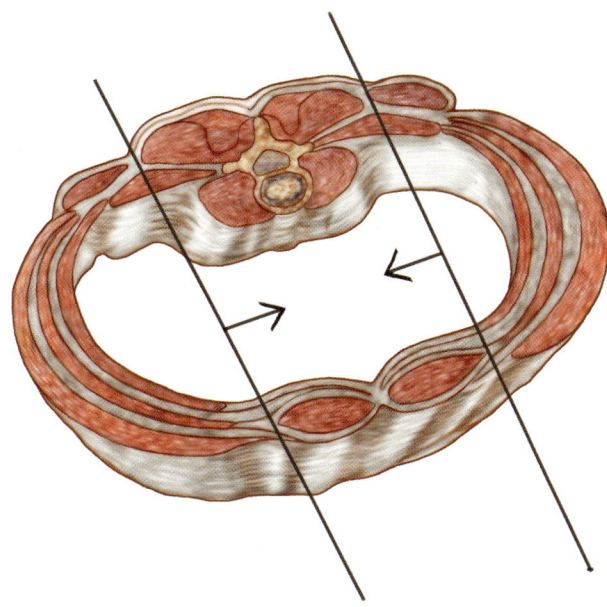

Fig. 5.13 The boards pass through the torso to join the eyelet lines at the back and front on each side.

1. Sit in a firm chair in neutral spine, so that your seat bones point straight down, and your feet are flat on the floor, with your legs slightly apart.
2. Put the edges of your hands on the front of one of the boards (Figs 5.12 and 5.13), laying them on the line from just beneath your bosom along the edge of the rectus abdominis (your 'six-pack') to the corner of your pubic bone. You are going to make that eyelet line come closer to your mid-line; think of rotating your body slightly to advance that board. Think too of advancing its back; do not let the board twist so that its back moves away from your spine.
3. Imagine that, if the third of your body to the outside of the board were full of stuffing, you could push that stuffing against the board in order to firm it up and move it over. Make sure that you do not collapse into a 'C' curve as you do this. Keep the side of your torso-box strong.
4. When you have done this, return to the neutral position, and notice that there is a symmetrical 'V' shape between your inner thighs, with the point of the 'V' towards the back of your underneath.
5. Repeat the movement you made above, and notice how it changes that 'V'. One of its arms should become longer and stronger (becoming a more distinct line of pull), with the other becoming shorter and fuzzier (a line of pull that has more slack). Most people find that the thigh on the side that they advanced and firmed up the board becomes the longer, stronger one. If you find the opposite, continue anyway, as repeating this on the other side should give you clarity.
6. As you reverse sides, put your hands on the front of the other board, and move that towards your mid-line. Think of slightly advancing the third that lies outside it, and pushing its stuffing against the board. It is probably already clear to you that this side is different – it is highly likely that you instinctively chose your more functional side to start with, so this second side is likely to feel less convincing.
7. Notice again how the 'V' shape of your thighs has changed. Practise on both sides and, having done both, you should be able to make sure that the arm of the 'V' on the side you are advancing becomes longer and stronger.
8. Once you have improved your awareness of each board, it is time for the punch-

line. Put on the board of the weaker, trickier side, and keep it in place as you then put the other board on too. Welcome to the experience of having 'both boards on'. This is reminiscent of two people both fighting to sit in the middle of the same barstool, but neither one can quite push the other one off!

9. Sit for a while in this position and *breathe*. You will probably find that you are bearing down stronger than ever, and that this is another 'stress position'. Notice how both of your thighs feel long and strong. Having this much tone in your thighs and torso is a completely unnatural way to sit, and you can work up quite a sweat whilst doing it!

Once you have become good at doing this exercise in a chair, do it whilst riding in walk, first on one rein and then on the other. Your horse may respond with some interesting steering, but let that be – it may reveal the underlying cause of steering issues you have not understood.

1. As you bring one board closer to the horse's mid-line, make sure that you encourage the other one to *fall away from it*, so that you are truly practising 'one side on/one side off'.
2. Give yourself enough time to find as much clarity as you can in each of four possible positions (i.e. right on/left off and left on/right off whilst riding on each rein).
3. In each position, rate the boards on a 0–10 scale. The ideal rating would be a 10 on the side that is on, and a 0 on the side that is off, but the numbers you give do not have to add to up 10.
4. Your pattern when riding in each direction could be very different, but most people soon realise that they do indeed ride with 'one side on/one side off' in one of the four options at least! So one version might feel like an exaggerated version of 'home' (and your rating might get close to 10–0), whilst its mirror image is an alien place that is hard to get into (and closer to 5–5).
5. In each position, aim to get a close as you can to 10–0. If one side is indeed much stronger than the other, you may find that it is *just as difficult to get the stronger side to let go as it is to get the weaker side to come on*. It is important to work on this, otherwise your strong side will continue to hijack any attempts you make to strengthen your weaker side. Not only does the stronger fascial line form a more distinct line of pull; the neurological pathways between that side and your brain may be much more developed. You need to learn how to send messages to the side you rarely 'talk to'.
6. Some people show other variations on the theme, perhaps with the top of one board being strong whilst the bottom of the other is strong. You may feel the front of the boards clearly but not the backs … there is no set response, but please rest assured that only two people have ever threatened to throw up as a result of doing this!
7. Do not put 'both boards on' until you have experimented for quite some time with one side on/one side off. Play with this when you hack out, and do it as part of your warm-up on every ride for several months, then return to it often.

With 'both boards on' (and both of those imaginary people fighting to sit on the same bar stool), your middle third has been squashed from both sides and made narrower. Indeed, it must become extremely narrow if we are to fulfil our aim of lining up each board and seat bone over the inside edge of the horse's long back muscle. Imagine your middle third being reduced to 'windpipe width' (see Figs. 5.8(b) page 104, 5.14(a)).

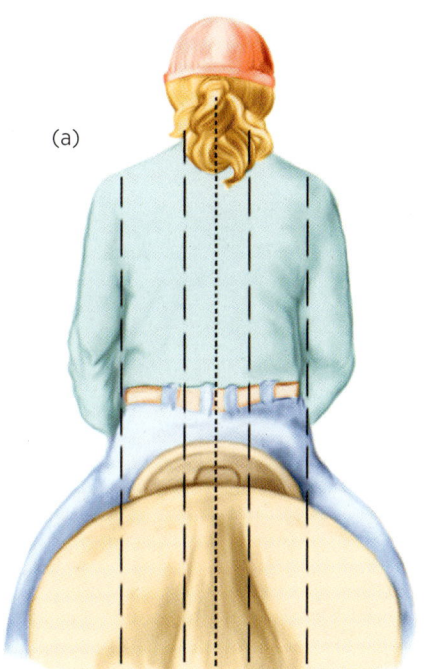
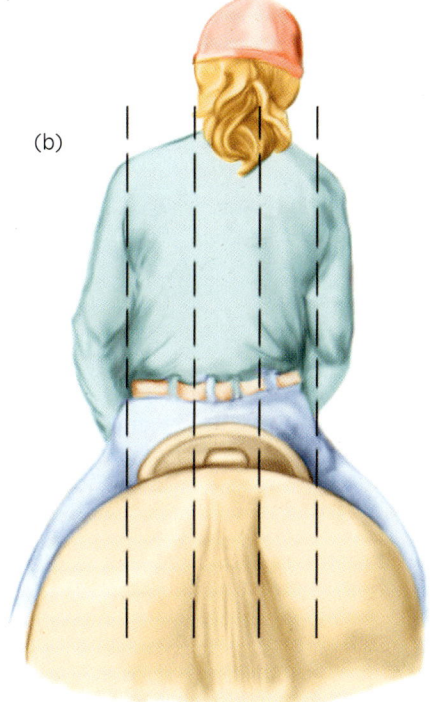

Fig. 5.14 In (a) the outer thirds of the rider's entire torso are stacked over the horse's long back muscle on each side. However, most riders show a 'C' curve similar to (b) when riding on a circle in one direction.

Your left and right thirds are now encased between the board and the outer edges of your torso-box. You might be shocked at the firmness this creates in your torso and thighs, especially if you have a history of relaxing – but it takes at least this much stabilisation if you are truly to ride both sides of your horse. Very few of us naturally have 'both boards on' when riding, and even really talented riders who do this exercise can usually find refinements that will significantly improve their steering.

Sitting this way makes the 'V' between your thighs much more symmetrical, with both of your thighs becoming longer and stronger. If one thigh keeps becoming looser when you ride, realise that the bottom of that board is also sliding away from the mid-line, taking your seat bone and thigh with it (see Fig. 5.8(b) page 104). You have unwittingly made a force from the opposite side of your body, which pushes that side off. To counteract this, keep repeating the exercises of Figs. 5.9, 5.10, 5.11 pages 105-106. I cannot overemphasise the value of these.

To understand the power of having 'both boards on', imagine using two boards to pick up autumn leaves in your garden. If you only had one board, you could chase the leaves across the garden in one direction – and potentially you could swop sides and chase them back again! But you could not pick them up. Similarly, the rider's boards can help to pick up the horse's back as well as steer him; but this is only possible when there is one on each side, and they are sufficiently close together.

Using the Outer Stability System to Solve the 'C' Curve Problem

I am hoping that when you ride on your more difficult rein, you can – for at least some of the time – keep equal weight on both seat bones, keeping them equally close to the horse's spine. Your aim is to keep your skewers pointing straight on, 'both boards on', both thighs on, and both knees level.

Having organised your mid-line as best we can, it is now time to think more about the outer stability system of the LLs. Most riding coaches attempt to solve the problem of the 'C' curve by lengthening the shorter line, perhaps by getting the rider to hold that arm up vertically. But this will not equalise the length and strength of the LLs, or the 'stuffing' (i.e. pressure) in each of the rider's outer thirds. I have

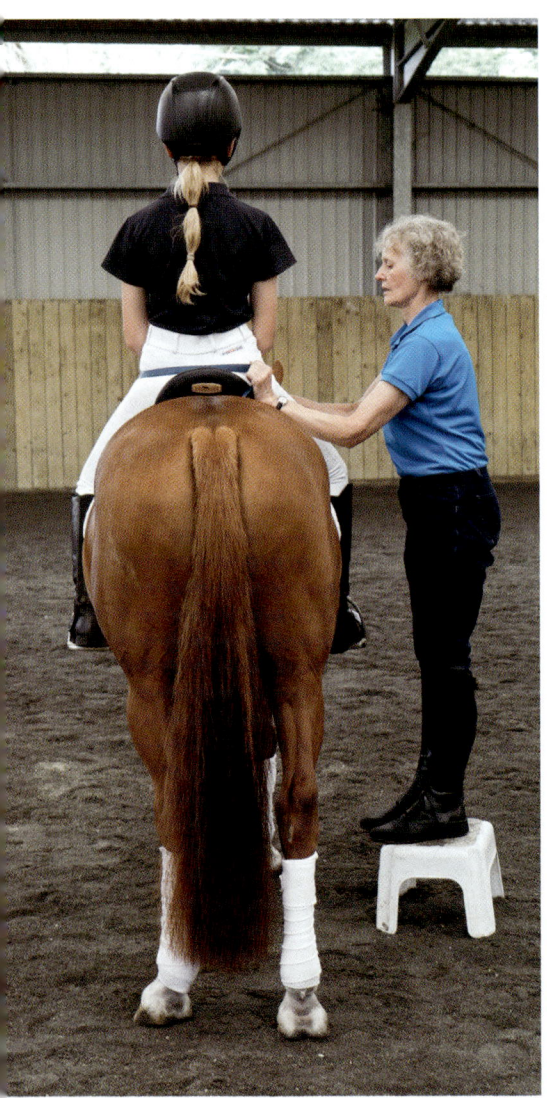
(a)

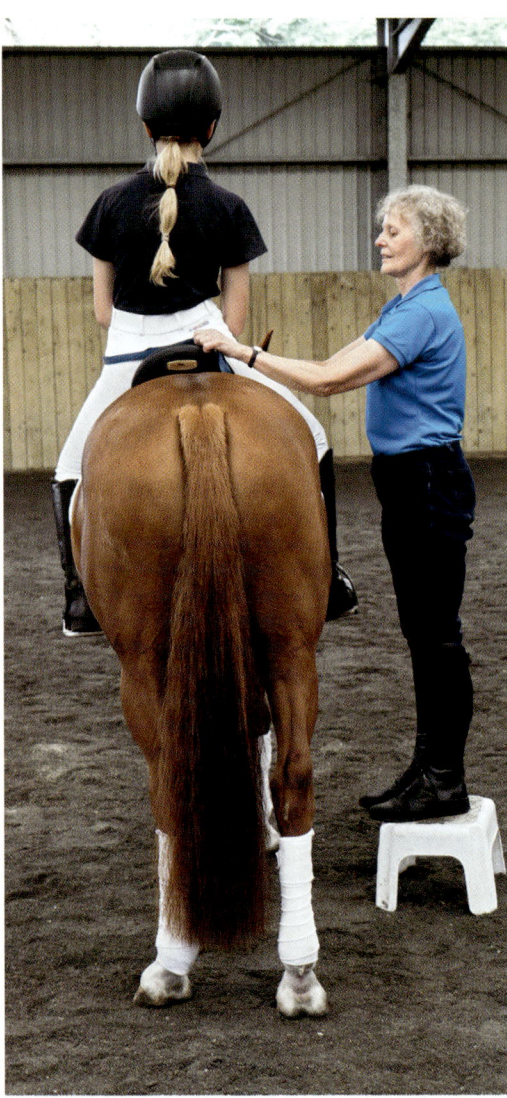
(b)

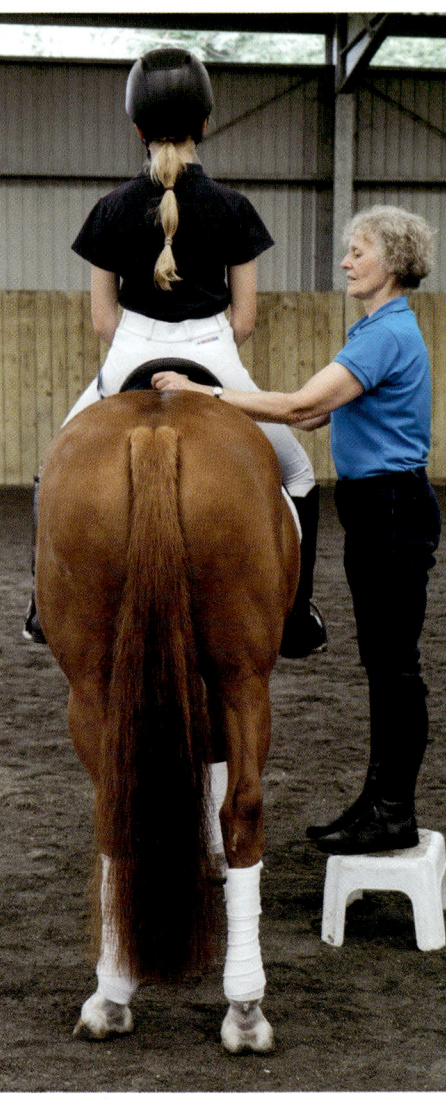
(c)

Fig. 5.15(a) A band around the rider's pelvis pulls her into a position with even weight on each seat bone and her mid-line over the mane. Remain aware of safety issues if you do this. When I let go as in (b), both rider and horse bulge to the side. Her challenge is to hold herself straight, and to feel and become able to reproduce the muscle work this involves. (c) Success! The rider held herself straight as I took the band away. She now knows which muscles have to work hard to maintain this.

had more success through showing riders how to *shorten and tone the longer line*, which, in turn, helps them address the problems of the shorter line. A friend or coach could help you as follows.

1. Halt so that when you walk on, your longer LL will be on the outside.
2. As coach, I stand on the rider's inside and put a band around the 'girlish' side of her pelvis. I then pull it towards me, and ask her how that has affected her outside seat bone and thigh. You will almost certainly feel that they have been drawn closer to the mid-line (see Fig. 5.15(a)).
3. Notice the weight on each seat bone, and aim to make it 50/50, with your chin and mid-line over the horse's mid-line.
4. When I let go, I invite the rider to let her pelvis fall into its natural place. The pelvis, thigh, and seat bone on the 'girlish' side all flop away from the mid-line, often quite dramatically – and often the horse's ribcage flops to that side too (see Fig. 5.15(b)). This is a brilliant demonstration of the force that acts within the rider's torso from the shorter side to the longer side. This exercise makes a great spectator sport, and will often evoke gasps from an audience!
5. Notice the weight on each seat bone, the contact of your inner thighs, the shape of your torso, and how your boards have deformed.
6. Several repetitions of this will give you the contrast between 'having it' and 'losing it', and your next task is to *hold your pelvis in place when your friend lets go*. This makes it clear that the muscles on the outside of your pelvis, and indeed your entire LL, must work extremely hard to stop it from sliding in to that 'girlish' place!
7. When riding on a circle, imagine a lunge rein wrapped around the outside of your pelvis and exerting the same force. Pressing one fist against the pommel may help this along by stopping your board, thigh and seat bone from sliding away from the horse's spine on the longer side (see Figs. 5.9, 5.10, 5.11 pages 105-106).
8. Having done this, it is now easier to influence the shorter side. Do the following exercise sitting in your chair now, and then as you walk your horse in the direction that puts your shorter LL on the inside.

Make sure that you are sitting (as best you can) with your skewers pointing straight on, 'both boards on', and both seat bones equally weighted. Then put your reins in your outside hand as you make a fist with your inside hand, and place it on your side between your ribs and your hips. Fill out your side against the resistance of your fist, feeling how this makes your side become stronger and less likely to crease (see Fig. 5.16 overleaf).

As you do this, you are increasing the pressure that your insides make against the wall of your LL. To increase it even more, put the side of your tongue against your upper molars on that side of your mouth, and then press your tongue against them. Feel how your insides now put even more pressure against the edge of your torso-box. Bizarre as it seems, imbalances in your 'bite' have a huge bearing on your LLs.

Fig. 5.16 Press your fist against your side in the space between your ribs and hips, and press your side against your fist. Also press the side of your tongue against your upper molars on that side.

You will need to repeat these fixes many times. Eventually, you may develop the opposite problem, or a new one entirely, but expect that to take months, or more likely years – if indeed it ever happens. Realise too that skilled bodywork, which seeks to rebalance your myofascial meridians (either skilled manual therapy, or yoga or Pilates-based training), will help you enormously in your quest to become a more symmetrical, effective rider.

Riders mostly change slowly, but it is highly likely that you will not need to ride your horse for very long before he begins to lengthen the same LL that you lengthen. He will also shorten the side that you shorten. Your challenge is to find out how to 'stuff' both you and him on both sides, as each successive moment unfolds, and for each new challenge you take on together as rider and horse.

Influencing the Horse Who Falls Out on a Circle

If the horse who jack-knifes and falls out on a circle were a stuffed toy horse, he would have a crease just in front of his inside shoulder-blade.

1. When the horse attempts to fall out on a circle, think of keeping his outside LL shorter, and his inside line longer. The requires his inside foreleg to lead the way to the inside, rather than his nose.
2. Think of preventing his nose from coming to the inside, by keeping it in front of the middle of his chest, or by making a slight counter-flexion, whereby he looks

to the outside. The horse needs to turn more like a bus, without hinging like an articulated lorry.

3. Imagine him like a stuffed toy horse, and think of putting more stuffing in front of his inside shoulder-blade. This may sound difficult, but relatively inexperienced riders can often make an immediate difference by invoking this image, as long as their thighs stay snug.
4. Set up each next quadrant of your circle by re-thinking about bringing his withers to the inside, so that he stays 'stuffed' as he steps to the inside. Keep repositioning yourself to make this possible. What do you lose when you lose it, and what do you get when you get it?
5. When you are most tempted to pull on the inside rein, you actually need to reaffirm how your outside aids are initiating the turn, and then *give the inside rein away*. Your aim is to make those outside aids into a wall that the horse cannot bulge through. This is only possible when your outside seat bone stays back and close to the horse's spine, carrying equal weight to the inside seat bone; without this, your outside hand and leg cannot do their job.
6. Also notice, what happens to your skewers, and to each of your boards?
7. What does your horse lose first? Since he has two unequal LLs, he almost certainly does not have two equal long back muscles. What kind of surface does he offer to each seat bone?
8. What does this teach you about your interaction with your horse? What is cause and what is effect?
9. Riding 10m circles in walk can really help you increase your awareness, and learn how to avoid falling into the traps set by the asymmetry in your LLs and his. How precisely can you steer his withers along the imaginary line?

When the Horse Falls In, or Wants to go Straight on

When the horse falls in on a circle, the easiest inroad is the following.

1. Realise that bringing his head to the outside will not straighten him – it will make him jack-knife and fall to the inside even more. Instead, you have to influence his ribcage and withers.
2. Imagine your inside thigh bone as a sci-fi weapon that can shoot laser beams out of your knee, or imagine it as a fire hose shooting water.
3. Ride squares or rectangles (perhaps in half the arena) rather than circles, and pick a point to aim to with your laser beams. This could even be in the far distance beyond the arena.
4. Just before the corner, let off that aim and then re-aim to the next point. You have to aim before the horse aims!
5. Continue like this until you can adapt the idea to riding circles, aiming out onto a bigger circle than the one you will actually ride on.

This tactic works much better than myopically pushing your inside thigh against a horse who is also pushing against you! If, instead, the horse has succeeded in pushing your inside thigh *off* the saddle, this tactic will keep it in place, especially when coupled with a resistance as in Figs. 5.9, 5.10, 5.11 pages 105-106. But it does not address the inevitable 'unstuffedness' of both of your outsides. However, many riders find that it works well enough to make a significant difference until they become ready to 'peel the onion' and reach that deeper layer of interaction.

Some horses brace their inside LL, making a strong contact against the bit, which gives them a distinct preference for going straight on! If this horse were a three-dimensional painting, he would be filled with rich, dark colour on his inside, and would be a washed-out watercolour on his outside. Often, the rider of such a horse will soon begin to match him, with all of her attention being drawn to that pull and that side of the horse. The rider thus meets strength with (her lesser) strength and compounds the problem – as her other side becomes a 'washed out watercolour'.

Whenever the horse falls in, and also if he is determined to go straight on, the rider has to learn *not to ignore the outside of his body*, with its lack of 'stuffing' (colour) and its attendant loopy rein. Riders rarely realise that this is at least 50 per cent of the overall problem. The horse cannot reshape himself to soften his inside without *filling out the outside and elongating the outside LL*. He will only reach into the outside rein (and lighten the inside one) by getting more stuffing (richer colour) into the outside of his ribcage. But he will not do that until *the rider* can get more stuffing (and richer colour) into her outside third, and also her outside arm, both of which will be as unstuffed (washed out) as the horse.

It is easier to influence the horse when his unstuffed side is on the inside. Once you can stuff him in front of his inside shoulder-blade, carefully change direction, realising that your aim is to keep the stuffing in place as that side of your horse becomes the outside. Ask yourself, 'Do I still have it? Now, now, now?' If you lose it, change rein to put the unstuffed side on the inside, and begin again. Shallow loops and then serpentines become good training exercises at this stage in your learning.

Realise too that the horse who falls in or out on the circle almost always speeds up or slows down as he does so. If the horse speeds up, he is likely to pull the rug out from under your feet so that you topple back. If your instinct is then to think 'Whoa!', you are likely to become the water-skier to his motorboat, and the plot thickens (see page 90). Conversely, if your horse slows down, you will probably instinctively let your body go soggy, and you will both lose the plot via the 'itsy-bitsy' trot (see page 90). If you are savvy enough to notice when one of these is about to happen, and you are 'with it' enough to *maintain the same trot*, steering will become much less of a drama.

The Lateral Lines in Your Legs

The fascial band that forms the LL in the outside thigh is no match for the stronger and more numerous muscles of the inside thigh. This partly explains the cramp you may experience when you do the traditional and tortuous exercise of holding your thighs out away from the saddle! Whilst this has less value than many believe, the rider does indeed need to find a balanced tension between the muscles of the inside thigh and the weaker LL.

A common manifestation of the inherent weakness in the LLs is 'noodly' lower legs that cannot be held still in sitting trot. It is as if the bottom of the lateral line cannot be anchored, but it will almost certainly be changes *higher up the line* that give the rider a sense of connection to – and control of – her feet.

1. The most effective way for a less-skilled rider to begin to 'switch on' the LLs begins with making sure that the thighs are rotated inwards. The inside thigh needs to be snugged in against the saddle, with an equal contact all the way from the corner of the pubic bone to the inside of the knee. The realignment of Fig. 3.6 pages 58-60, is the ideal way to create this.
2. Push your knee and/or heel out against a resistance, which can be offered by a friend or coach. Do not take your thigh and/or knee off the saddle as you do this, and be sure that you do not let your ankles roll outwards so that your weight goes to the outside of your foot. Think: thighs rotated in, knees in, toes in, heels away from the horse's side (see Fig. 5.17).
3. After you have practised with a resistance on your ankle, have your friend add a resistance on the fist of your outstretched arm as shown in Fig. 5.18. This

Fig. 5.17 Pushing the heel out against a resistance helps to position the lower leg and strengthens the outside of it.

Fig. 5.18 Pushing both the heel and the outstretched arm into a resistance activates the entire Lateral Line.

Fig. 5.19 The coach runs the knuckles of two fingers down the Lateral Line in the thigh and calf, which makes it feel longer and stronger. A short Lateral Line makes the rider weight her big toe; this exercise helps her spread weight more evenly across her foot. (This exercise is Included with kind permission of Julie Houghton of Equestrian Balancing Therapy.[5])

also strengthens the 'Xs' on the side of the torso, firming up the entire LL. Try pushing your tongue against your molars at the same time. Doing this exercise on both sides of the body usually makes the discrepancy between the length and strength of each line very obvious. Repeat the exercise on your weaker side.

4. If your weight rolls easily to the inside of your foot so that an observer can see the soles of your boots when you ride, have your coach take hold of your breeches on the outer upper thigh. Ask the coach to run the knuckles of her first and second fingers slowly down the line of the iliotibial tract, and over your boots down the outside of your calf to your ankle. Repeat this three times. It is a great feeling, and helps to elongate the LL (see Fig. 5.19).

A Helpful Off-horse Exercise[6]

1. Lie on the floor with your longer LL by a wall. Curve the rest of that LL slightly away from the wall (see Fig. 5.20).
2. Resist your shoulder and foot against the wall. Each resistance should last for six seconds, with a five-second break in between.
3. Cross the ankle furthest from the wall over the other one, and simultaneously resist the top ankle away from the wall, pushing it against the bottom ankle,

Fig. 5.20 Lie with your longer side slightly curved away from a wall. Resist your shoulder and ankle against the wall. Cross the ankle of the leg further from the wall over the other one, and simultaneously resist this ankle away from the wall. You can also resist that same side against your fist, and press your tongue against your upper molars! (This exercise has been adapted from Muscle Activation Techniques, devised by Greg Roskopf.)[6]

which is still resisting into the wall. This will expand the 'soggy' side of your ribcage that could so easily make creases.

4. Press your fist against that side, and resist your side into it.
5. Think of the laces of your LLs; they are looser on your longer side and tighter on your shorter side. On the longer side that is beside the wall, they need to pull your front and back eyelet lines closer together, lacing them tighter around the edges of your torso-box. As you elongate the other side, think of the eyelet line on your front being pulled up and away from the one on your back.
6. Six repetitions done several times a day will begin to yield results, but expect it to take some time to become markedly more symmetrical.

An Overview: the 'Double Yellow Lines'

Ideally, the rider's boards stabilise the horse on each side of his mid-line, since the narrowness of 'both boards on' positions them over the inside edge of the horse's long back muscles (i.e. the outer edge of the gullet of the saddle). Concurrently, well-toned, well-aligned LLs (that form the edges of the torso-box) set limits that the horse is not allowed to bulge through. In theory, he does not even try to bulge through them – but in practice, untrained bodies, both human and equine, inevitably have distortions here and usually they amplify each others' issues (see Fig. 5.15(b) page 112).

Anatomically, if the horse had boards that were analogous to yours, they would connect down through him, joining the eyelet lines on the outside of his long back muscles to the eyelet lines on the outside of his abdominal muscles. But we are going to think right now about the *inner edges* of his Superficial Back and Front Lines (which lie under the outer edge of the gullet of the saddle). To give you an overview of how the intermediate stability system and the LLs function in relation to your own and the horse's Superficial Back and Front Lines, think of the double yellow lines that indicate 'no passing' on American roads, and 'no parking' on British ones. Imagine similar lines going from the heels of the horse's hind legs, over each hock and hamstring on that same side of his croup, along his long back muscles under the panels of the saddle, up each side of his crest to each ear. These are his Superficial Back Line.

When these muscle chains function well on both sides and the horse is working 'over his back', each long back muscle gives the rider a firm surface to sit on, and the rider's pelvis fits over each side as shown in Fig. 5.8(b) page 104. I find it helpful to think of the horse's long back muscles like four- or five-inch wide strips of rubber – rather like the rubber that forms the conveyor belt at a supermarket checkout, but with slightly more give.

Think of the 'double yellow lines' also going up your back, over your shoulders and down your front. Your aim is to stack your yellow lines over the horse's yellow lines – all the way up to your imaginary epaulets (as on a military or doorman's uniform), which mark the top of the outer thirds of your torso. Clear boards and firm edges to your torso-box make this possible – and add enormously to your horse's straightness and 'steerability', enhancing the tensegrity of the rider-horse system. Your influence becomes even greater if you can 'think' a connection down through the horse's boards to the eyelet lines that form the outer edges of his Superficial Front Line.

With time, dedication, and the skill of progressively noticing more subtle distinctions, it is possible to develop enough resilient strength in the outer and intermediate stability systems to transmit forces through your own and the horse's body in the direction you want. More commonly, forces are transmitted in the direction that *he* wants, or in the direction determined by a combination of both of your asymmetries with the addition – on a circle – of the centrifugal force. It takes precision and power to overcome these obstacles. But when muscle-chains align, your work can look powerful and easy, and you have a wonderful connection with your horse.

CHAPTER 5 NOTES

1. Clayton, Hilary, 2015, personal communication (2015)

2. Bennet, Dr Deb, 'Lessons from Woody', on www.equinestudies.org

3. McLean, Andrew, *Academic Horse Training*, Equitation Science International (2008): www.esi-education.com
   McLean, Andrew and McGreevy, Paul, *Equitation Science*, Wiley-Blackwell (2010)
   McLean, Andrew, 'Thinking about horses, Part 2', www.horsemagazine.com/thm/2015/01/andrew-mclean-thinking-about-horses/

4. McLean, Andrew, 'A fresh look at the training scale', www.horsemagazine.com/thm/.../a-fresh-look-at-the-training-scale-with-andrew-mclean

5. Houghton, Julie, Equestrian Balance Therapy: www.equinetherapyservices.co.uk

6. This exercise has been adapted from Muscle Activation Techniques, devised by Greg Roskopf, see www.muscleactivation.com

PART 3: FUNCTIONAL LINES AND ARM LINES – PUSHING THE HANDS FORWARD

CHAPTER 6

THE FUNCTIONAL LINES

The lines we have considered so far traverse the front, back, and sides of the torso and legs. But in order to trot and canter, a horse needs at least some of his muscles to cross the mid-line, joining the contralateral girdles, i.e. linking the left hind leg to the right foreleg, and vice versa. We need these connections too, so now we meet the Functional Lines (FLs), which involve superficial muscles and facial sheets that function more to coordinate movement than to maintain stability. This makes them easier to work with, and that work yields great rewards.

In humans, the front and back FLs consist of four separate lines that form 'Xs' on the front and back of the body. They are activated in every walk step, since with each forward step of your right thigh, your left shoulder advances as a counterbalance (and vice versa). These two 'Xs' across our front and back coordinate the two 'watch-springs' of the thorax and the pelvis as they wind and unwind in our walking and running.

The FLs get their name from the vital function they play in magnifying diagonal movements, as in kayaking and most sports movements, including a tennis serve, a javelin throw, a baseball pitch, or a cricket bowl. The FLs can also be used in reverse – the arms stabilising the legs, as in hurdling or a football kick.

For the horse, as we will see, it is less about an actual spiral differentiation in the trunk, and more about coordinating every trot step, as opposite limbs move together. In him, our four lines become two: the front and back FLs that begin just above each elbow form a complete loop across back and belly, from here to the opposite stifle. His torso is wrapped diagonally, as if with two parcel strings.

For riders, the significance of our FLs lies in the benefits of centring and equalising each 'X', and in the way that the lines fix the base of support for the opposite arm. The FLs, in effect, continue the Arm Lines (which we will meet in the next chapter) diagonally across the torso to the opposite side of the pelvis and the opposite leg, helping to stabilise the arm and also that diagonal. Unlike the runner or walker, the rider does not want a lot of 'watch-spring' action whilst riding – like the horse, the rider wants to maintain adaptability whilst keeping the rotation in the trunk to a minimum.

Let us look at the FLs in more detail (Fig. 6.1). The Back Functional Line (BFL) (Fig. 6.1(b)) begins on the arm bone close to the shoulder, and continues via the tendon that you can easily feel if you grasp the back of the armpit (see Fig. 3.12 page 69). The line then travels through the latissimus dorsi muscle. This is the big triangular muscle that covers each side of the back, from the back of the armpit to the top of the pelvis and sacrum on the same side. When you ride, pulling down the muscle and tendon on each side gives you stability. In birds and bats, the 'lats' power the wing.

The BFL follows a path through the lats that leads to the mid-line at the top of the pelvis. Here it crosses the mid-line of the body, and continues through fibres near the bottom of the big gluteus maximus muscle (commonly known as 'glutes'), that forms your backside. The line passes under the muscles of the outer thigh (that form the Lateral Line), and attaches into the outside/back of the thigh bone. It then continues via the outermost muscle of the quads to the outside of the thigh and knee. Here it attaches to the top front of the tibia, the bigger bone in the calf.

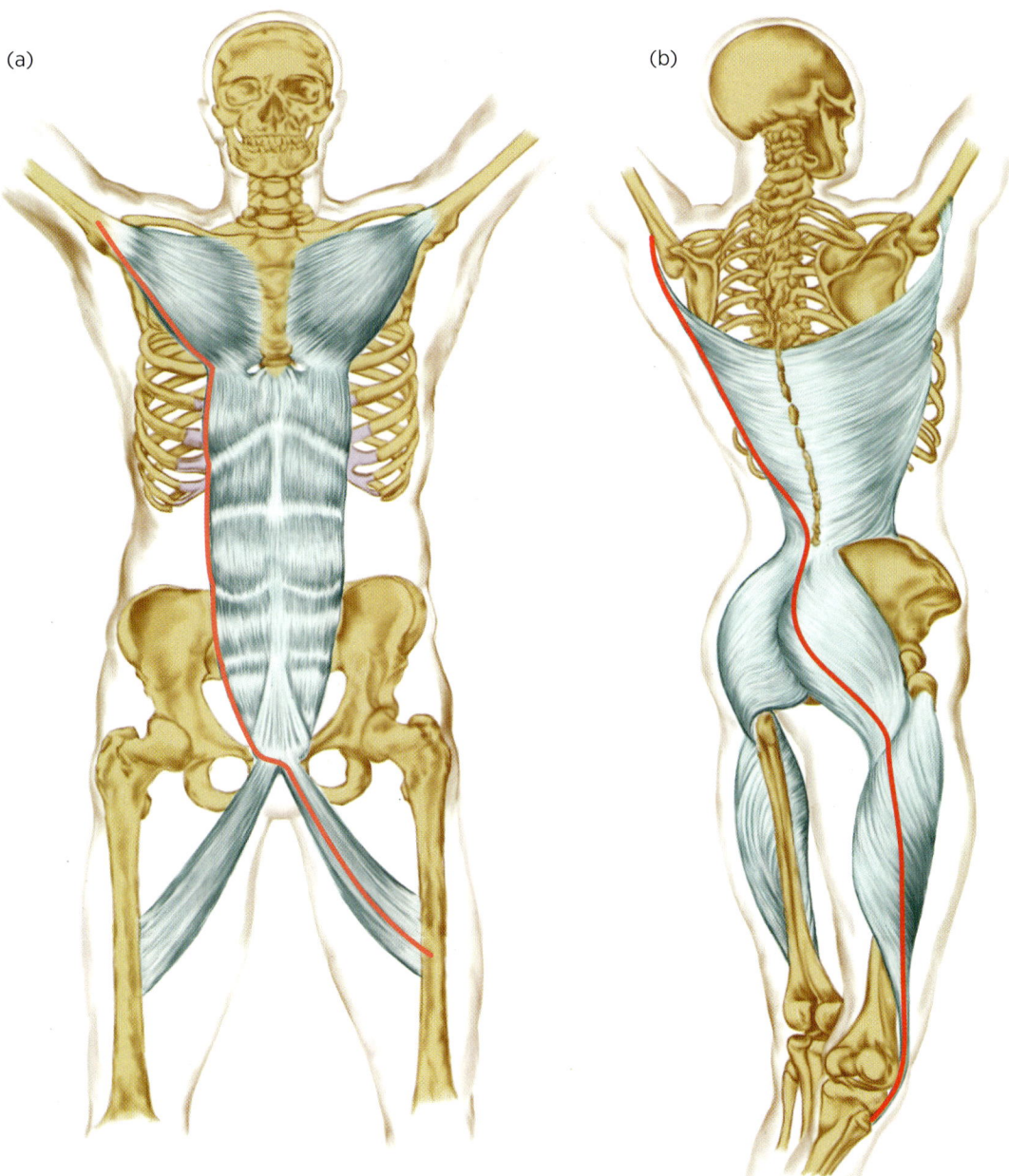

Fig. 6.1 The Front Functional Lines (a) and Back Functional Lines (b) form 'Xs' on the back and front of the body. It is easier for riders to think of the front 'X' taking a more direct line from one armpit to the opposite side of the pubic bone.

Fig. 6.2 Reach one arm across your front, grab the front of your armpit, and pull down your shoulder and elbow. Feel for the tendon that attaches the muscles of your pecs into your upper arm. The shoulders and this tendon should be pulled down as you ride.

Although it is not one of Myers' Anatomy Trains, from here you may imagine the line continuing to the outside of your foot. This imagined extension of the FLs along the outside of the calves, via the Lateral Line, can add significantly to your ability to stabilise your lower legs. What joy!

The Front Functional Line (FFL, Fig. 6.1(a) previous page) begins on the front of the arm bone near the shoulder, and continues via the tendon that you can easily feel in the front of your armpit. Reach one arm across your chest, grasp the front of your armpit, and feel for the tendon as you pull down your shoulder and elbow. Once again, this tendon should be pulled down as you ride.

The FFL continues into the lower edge of the pectoral muscle, which forms the bulk of the chest in a well-muscled man. The fascia beneath this also includes the smaller pectoralis minor muscle, which inserts into the fifth rib along the line of the board (see previous chapter). This has significance that will become clear in Chapter 7 on the Arm Lines.) The fibres of the pectoralis major run diagonally to its insertion on the outer edge of the big rectus abdominis muscle, which we know as the 'six-pack'. As you can see in Fig. 6.1(a) page 123, it is really a ten-pack, with its highest 'pack' above the sternum. From this point by the fifth and sixth ribs, the line continues along the edge of the rectus abdominis on that same side. From the waist downwards, the outer edge of this muscle curves gently towards the outer edge of the pubic bone.

Fascial fibres connect through and over the cartilage between the two sides of the pubic bone (the pubic symphysis) to the strong, obvious tendon at the top inside thigh (the one that gets sore if you ride wearing the wrong underwear!). This belongs to the muscle adductor longus, which inserts into the inside of the thigh bone about halfway along its length, about where the thigh lies over the stirrup bar of the saddle.

In practice it works well to think of the arms of the 'X' forming a straight line from one armpit to the opposite groin, putting the cross of the 'X 'a bit higher up than its actual position, which is your pubic bone (Fig. 6.1(a) page 123). The FFLs insert into the thigh bone about halfway down it, but once you are good at working with the them, the FFLs become even more powerful by extending them in your imagination down the inside of your thighs and calves to the insides of your feet.

Working with the Functional Lines

To avoid confusion we name each Functional Line for the shoulder it starts from, so the right FFL includes the front of the *right* armpit and the *left* hip.

1. Which FL are you intuitively drawn to as you start this exploration? This will probably be the one that is most robust.
2. Track its path on your body, beginning at whichever end you choose, and feel its tone – would it be toned enough to be plucked like a guitar string, or is it too 'soggy' to 'play a note'? Is the tone even from one shoulder to the opposite hip? Does the line cross your mid-line, or does it fade out instead?
3. Clarify with your felt-sense where the line inserts into your thigh. The weakness of at least one FFL is likely to contribute to the inside thigh that stays further from the saddle than it should.
4. To remedy this, put the first two fingers of the hand on that side just inside the pommel, and pull outwards, as shown in Fig. 5.10 page 106. Feel how this puts the tendon at the corner of your pubic bone more against the saddle. You are reducing the 'wriggle room' that you give the horse's withers: think of your thighs as both pieces of bread in a horse sandwich!
5. At least one of the BFLs will almost certainly become very indistinct as it curves around your backside and into your outside thigh.

Fig. 6.3 Put your arm behind you, grasp the opposite side of the cantle, and pull on it to firm up the Back Functional Line as it passes through your glutes to your outer thigh. Your left arm will firm up your left glute, which is part of the right Back Functional Line.

6. To remedy this, reach the hand from the opposite side across the back of the saddle and pull on the side of the seat just behind your backside (see Fig. 6.3). Feel the BFL firm up and shorten, increasing the tone in that corner of your backside. Can you feel just how wobbly you have been up to this point?
7. Once the first line is clear to you, add the one that next appeals to you; but do not be in a hurry to do this, or to go beyond working with two lines.

Whichever two diagonals you chose to begin with, be they both fronts, both backs, or the front and back lines that originate in the same shoulder or (for those with rotations in the spine) even opposite shoulders, expect them to have some aberrations. One line may well be strong and tangible whilst the other is weak. One may function well at the thigh end, but not at the shoulder end, whilst the other has the opposite pattern. Expect to discover that at least one shoulder lives life in a place that you now realise is far from ideal – and realise how shoulder mal-position can affect your seat! Changing this could be a very significant improvement in your ability to both stack up your torso and steer your horse.

Working with two of the FLs may keep your brain occupied for quite some time, but monitoring all four eventually becomes possible. Once they become part of your unconscious competence, expect that you will revisit them at times as new insights emerge.

Off-horse Fixes

Timing the Functional Lines
The timing of how the FLs work is highly significant.

1. Lie on the floor, on either your front or back, depending on which lines you want to work. Lift the contralateral limbs off the floor – e.g. your left arm and right leg. Listen to the movement; the chances are that you lift one limb a fraction of a second before the other. Work to lift them exactly together, and you will feel the middle of the line as it crosses your torso.
2. To create more strength in the line, you can add weights to your lift – small kettlebells work well – or have someone hold your limbs so you can push against the resistance. This works well for the FFLs, but is essential exercise for the BFLs – lie on your belly and lift each line of the 'X' into the air at exactly the same time to feel your thoracolumbar fascia in sharp relief.

The following off-horse exercise helps to shorten the longer BFL, and to develop your awareness of it.

1. Lie on your back, either on the floor or on a firm mattress, with your legs long.
2. Pull your greater trochanters (see Fig. 5.2 page 93) down towards the floor, and in towards each other. This will make your pelvis lift slightly as your gluteal muscles firm up.
3. Very gradually, let those muscles go. Can you do this smoothly and evenly, with both sides in unison, or are there some 'clunks' in one or both sides? Do both sides reach their maximum of 'let go-ness' at the same time?
4. If not, notice how much extra letting go is possible on the longer, weaker side. This will also be the side of the longer Lateral Line.
5. Repeat the exercise, and stop the longer, looser side from doing the extra 'letting go' it would like to do. Hold it on a level of tone that matches the side that is already fully let go.
6. After a while, let the weaker side go completely, and then firm it up again to the level that matches the other side. Hold it there for a few seconds, and then begin the exercise again by firming up both sides.
7. Do a few repetitions of this every day, perhaps in bed before you go to sleep. It will take less time than you expect to reduce the 'clunks' and increase the tone in your longer, weaker BFL.

The value of this is huge, since the longer BFL allows the ribcage to wander away from your pelvis, giving your backside and seat bone the leeway to slide away from the mid-line. The centres of the 'Xs' then migrate away from your centre, and the 'C' curve of your longer Lateral Line lengthens the outside of your body. You are highly likely to rotate inwards, so that your torso faces the inside of the circle … and you already know the rest of the story!

Centring the Cross of the 'X'

Realise how the FLs relate to your Lateral Lines and any lateral 'C' curve in your torso. Unequal FLs make the front and back 'Xs' asymmetrical, shifting their centres away from the mid-line of the body.

1. As you sit reading this, make a 'C' curve with your torso. Where are the centres of the 'Xs' of the FLs relative to your spine? Feel how their centres have moved off the mid-line.
2. Imagine a hand that could pull on your spine to line it up with the centre of the back 'X'. Put the centres of both 'Xs' on your mid-line, with your spine vertical and your 'box' symmetrical. What does it do to the 'stuffing' and pressure in each side of your torso?

It works well to think of the FFLs and BFLs in your torso all forming a giant cavalletto 'X', with square corners that fit into your backside and shoulders (Fig. 6.4).

Fig. 6.4 Imagine a giant cavalletto 'X' inside you, with square corners that fit into your shoulders and backside.

The Functional Lines in Your Legs

The net result of continuing both lines down to your feet – the front 'Xs' down the inside of the shin and the back 'Xs' down the outside – is a dramatic increase in your ability to keep your lower legs still, and to give leg aids in the way you want to.

1. Do your lower legs lie against your horse's sides, or away from them? The resistance of heels out (see Fig. 5.17 page 117) helps to keep your calves away from him (as your thighs stay against the saddle). This gives him a clear contrast between when you do and do not give a leg aid, which helps to ensure that your 'kick' retains its meaning.
2. Is your leg aid more like a slap, or a nudge? A slap (whether light or more hefty) is a quick touch that rebounds. (The movement of a kick continues in the same direction after meeting its target, but 'kick' is still a relatively good term.) A slap (kick) is much more effective than a nudge, which is relatively prolonged, and has more force – tempting most riders to add some contortions.
3. As you kick do your calves go 'soggy'? Do your legs aim inwards or backwards? Do you turn your toes out, lift your heels, and/or contort your feet in any other way? Which leg is least effective, and what goes wrong?
4. Your aim is to keep your heels away from your horse's sides until you want to kick, and then to maintain the firmness of your calf as you do so. Kick *inwards*, and imagine your calves and feet like a solid wooden boot tree within your boot.

The connection of the FLs into your lower legs might make you feel as if you are 'stretching your legs down', and give you the feeling that the inside and outside of your foot are connected to the front and back of your opposite shoulder, respectively. Gaining more control of the lower legs is a big issue for many riders, and including the FLs in your body awareness can be a significant step towards it!

Myers has recently added another line to this set, running from the posterior part of the armpit down the sides on the front edge of the lat (which we felt before) down to the lower outer ribs, over the front of the hip, and via the sartorius muscle to the inside of the knee. This muscle follows the seam line of an old-fashioned pair of breeches, sweeping across the front of the thigh. Via this Ipsilateral Functional Line (IFL) we can feel the connection between the leg and the *same* shoulder, as well as across to the other one. Thus the IFL connects the 'box' of your torso to the area where the inside of your lower thigh and knee contact the saddle.

The Functional Lines in Your Horse

As I said in the beginning, the FLs in the horse operate a bit differently from those in the rider. Our torso is designed to rotate through our waist between the ribs and pelvis, whereas most quadrupeds, like your horse, use little rotation in the trunk. Throughout their various gait rhythms they have to counterbalance through the coordinated motion of the limbs, which is organised primarily by the central pattern generators within the spinal cord.[1]

Within the diagonal 'parcel strings' of the FLs, imagine a large 'X' painted on the horse's back, with you sitting right around its crossing point (see Fig. 6.5(a)). Its arms begin just above his elbow on each foreleg, and via his lats they cross his spine in the large sheet of thoracolumbar fascia. This stretches like a saddle cloth directly under and behind the saddle itself, lying on top of the two rails of muscle that form the Superficial Back Line.

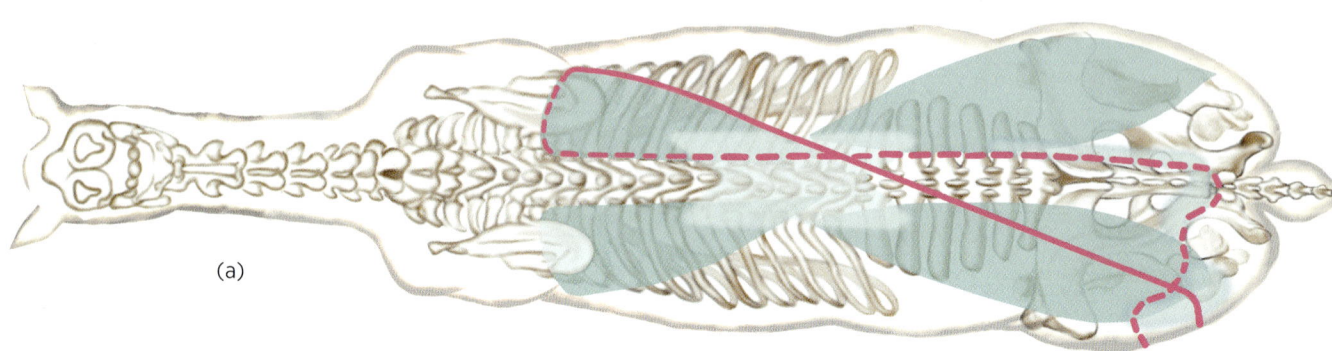

(a)

Figs. 6.5(a), (b). The Functional Lines in the horse, seen from the top and side, with the right Functional Line highlighted in pink. The dotted lines indicate that the line is on the far side of the body, or the underneath of the body. The lines cross, forming 'Xs' on the horse's back, and by his pubic bone, as they also do in us. In the horse the front and back Functional Lines that follow the same diagonal meld together to make a complete ring (that is not quite so simple as 'parcel strings'!). Fig. 1.1(c), Chapter 1, shows fascial fibres crossing near the pubic bone from the inside thigh to the abdominal muscles on the opposite side.

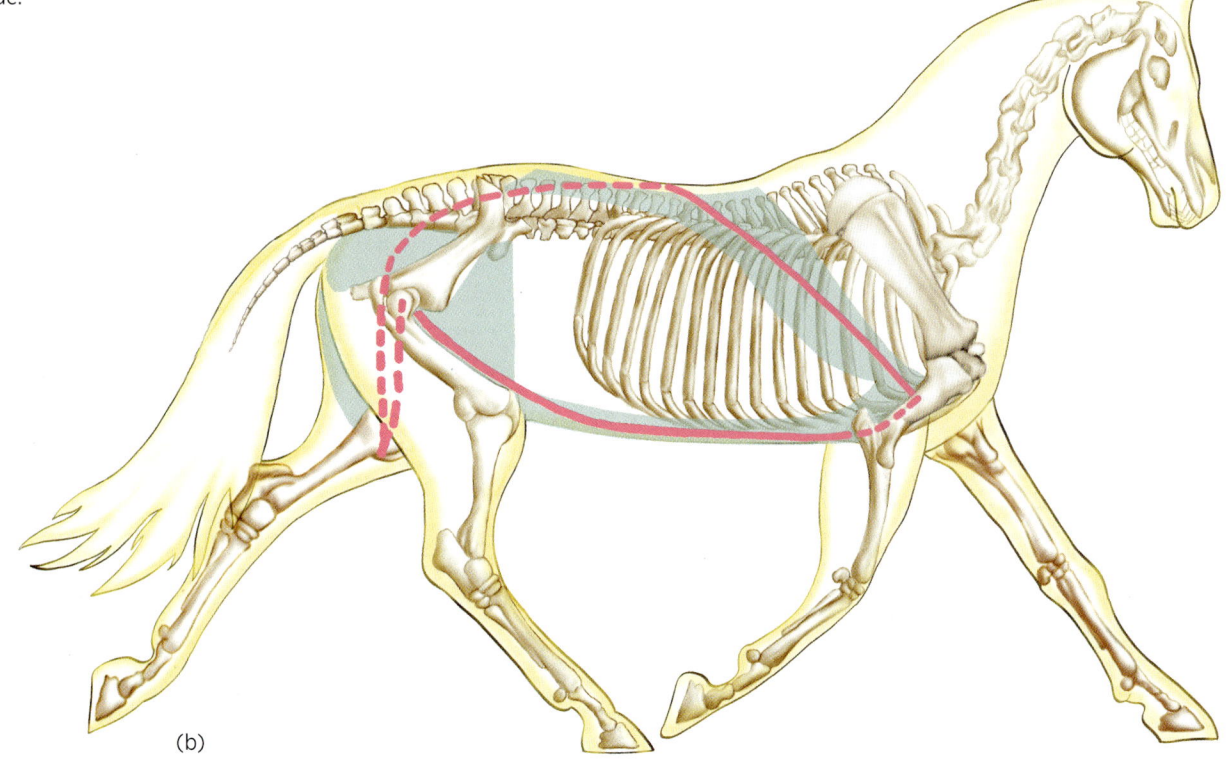

(b)

The back arms of the 'X' continue into the outside of each opposite hindquarter. They reach down to the stifle, passing under the patella (his kneecap) to the inside of his thigh, and then back up it to his pubic bone (see Fig. 6.5(b)). Like our FFLs, the lines then cross to the opposite side, and continue along his abdominal muscles to meet his pectoral muscles. The fibres of these meld with his lats above his elbow – making a complete diagonal ring.

To think more about the FLs on the horse's stomach, make a human/horse comparison by putting your arms in front of you. Realise how the horse's shoulder joints are in against his torso, whereas ours are to the side – so his pectoral muscles are shorter and denser than ours. They connect into his abdominal layers of fascia, which have to be much stronger than ours to hold up his massive digestive system. Likewise at the hips, his adductor muscles are shorter and stronger, being flexed and held close to the body.

Like the 'X' on the horse's back, the 'X' under his belly is there not to create more rotation (as we do when we walk or throw) but to *prevent* trunk rotation by counterbalancing the contralateral girdles. Whilst this stabilisation is their primary role, the FLs also flex and extend the horse's back, depending on their relative lengths in his back and belly.

I see a small percentage of horses in whom I suspect this is particularly significant. They take each step rather like a ballet dancer doing an arabesque: their forelegs advance extravagantly, whilst their hind legs extend far out behind them – their elbows and stifles come apart as they do when a horse urinates. When you are riding this type of horse (who inverts somewhat differently from the majority) it can help to think of his elbows staying closer to his chest, so that they cannot 'run away' from the rest of him.

Since the primarily role of the FLs is to stabilise the diagonal motion of trot, I teach riders to work with the Superficial Back Lines long before we talk about the FLs. Some of the force that is transmitted from each hind leg is directed via the Superficial Back Line 'over the back' to the poll on the same side, whilst some passes via the FLs to the opposite foreleg.

In time, we will know much more about this force transmission. Meanwhile, Professor Emerita Hilary Clayton is convinced that a significant underlying (and frequently overlooked) factor is the horse's pelvic stability, since, without this, forces cannot be transferred from the hind limbs to the spinal skeleton.[2] In practice, the rider's developing awareness of her own and her horse's Superficial Back Line (and Superficial Front Line) improves many facets of ability to sit and steer well, as it concurrently improves the horse's carriage. The diagonals of the horse's FLs often take care of themselves if we pay attention to the longer longitudinal lines.

Some Thought Experiments with the Horse's Functional Lines

In both humans and horses, the latissimus is the only muscle that attaches the arm (or foreleg) to the pelvis. For an experienced rider who has discovered how powerful focused intention can be, thinking about this can help to lighten the horse's forehand. The rider's focus helps the horse to generate a fascial pull within his lats from his forelegs towards his lumbar and pelvic area, shifting some of his weight from his forelegs to his pelvis and hind legs. Imagine his lats and skin being pulled backwards over his bones, creating overall tone and, we might say, an impressive tensegrity.

Thought experiments can also help to address asymmetry in the horse's FLs. Since they are the diagonals in the rectangular box of the horse's torso, differences in their length means that the rectangle is not truly rectangular. (The same is true of us.) The discrepancy usually causes the horse to 'crab' like an aeroplane in a crosswind (probably towards home or his friend!). His torso moves on a slant, with his

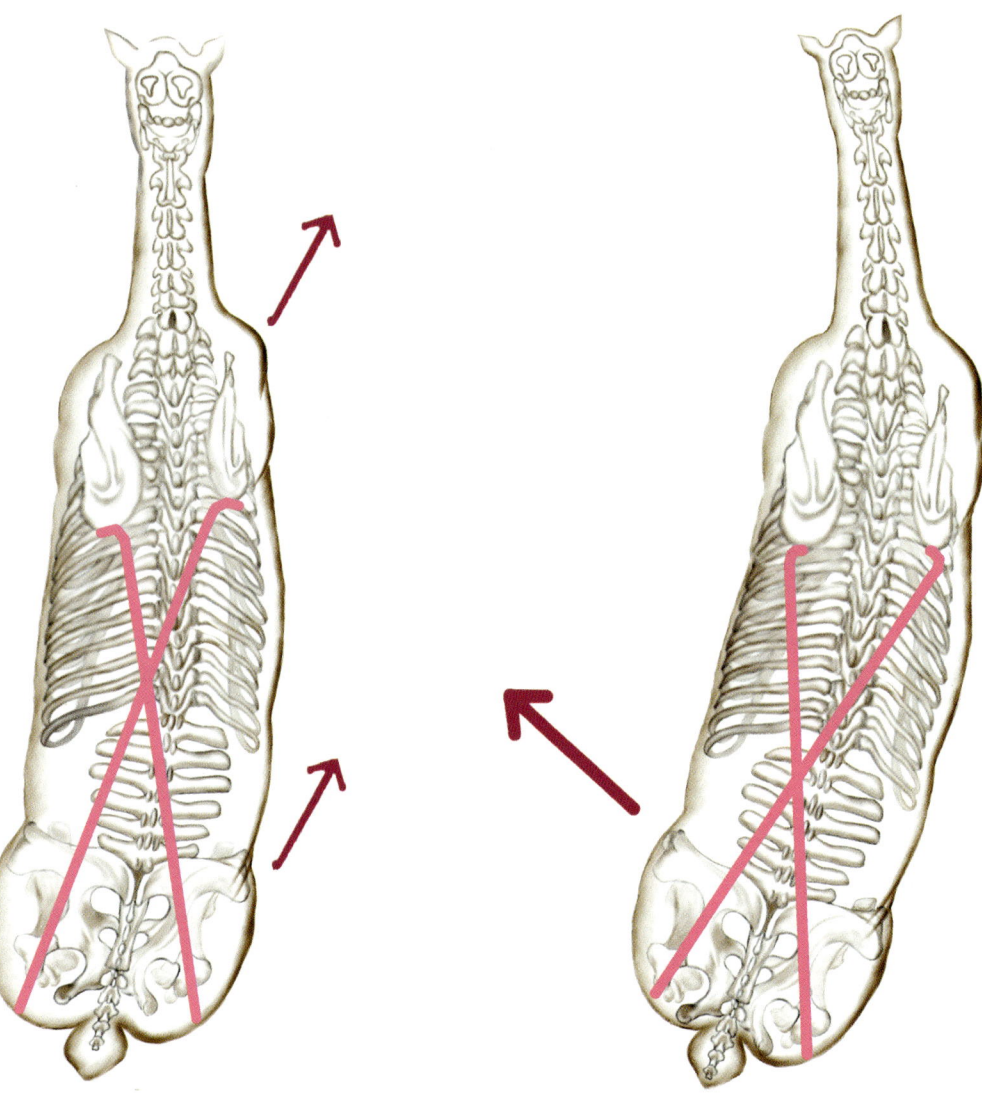

Fig. 6.6 The horse with uneven Functional Lines moves on a slant, with the rectangle of his torso displaced so that his hindquarters move to one side of his shoulders. He may or may not jack-knife at the same time (see Fig. 5.5(c) page 97.

Fig. 6.7 Both the Functional Lines and the Lateral Lines can cause the horse to hold his quarters to one side of his shoulders, so his torso is not a rectangle. This horse will not easily bring his shoulders to the left of his quarters, so circles and shoulder-in left are difficult.

fore and hind legs taking different tracks (see Fig. 6.6). As this happens, he may or may not jack-knife, as in Fig. 5.5(c) page 97, in the Chapter on the Lateral Lines.

When you ride, imagine the lines of the horse's FLs and where they would cross on his back. Is this point on his spine, or to one side of it? Imagine that you can adjust the lines to cross on his mid-line, and arrange yourself right over the cross of the 'X'. This area can become a place where energy gets blocked, but focusing on it may be enough to make a difference.

Fig. 6.7 shows how, on the horse's 'soft side', his unequal FLs can join his uneven Lateral Lines (and Superficial Back and Front Lines) to bring his hindquarter and hind leg to the side as his body bends, especially behind the saddle. The rider struggles to bring the horse's shoulders in front of, or to the inside of, his quarters. Both riding a circle and riding shoulder-in to that side is difficult – the horse just wants to offer you travers, and will distort your torso-box into 'travers position', with its inside too advanced. This gives him your ongoing permission to move with his quarters to the inside!

In the opposite direction he may well fall in on a circle, and the rider's attempts at shoulder-in will easily morph into a leg-yield with too much angle. I confess that, historically, all of the horses I rode would have preferred to put their hindquarters to the right – and I have colleagues and pupils those horses show the opposite (as well as the same) pattern. But regardless of whether the pattern originates with the horse, or with the rider, the latter is the only partner with an interest in changing it – and the challenge for the rider is to first even out herself, and then her horse. All of the lines we have met so far are involved, and the proof of the pudding is that *the rider* is the one who organises her own and her horse's 'double yellow lines', and determines their positioning in the various movements.

CHAPTER 6 NOTES

1. McLean, Andrew 'Thinking about horses, Part 2', www.horsemagazine.com/thm/2015/01/andrew-mclean-thinking-about-horses/
2. Clayton, Hilary, private correspondence, 2016

CHAPTER 7

THE ARM LINES

The Human Arm Lines

The detail of the myofascial lines in the arm is quite complicated, but fortunately we as riders can glean all we need to know from a broad understanding of how they operate.

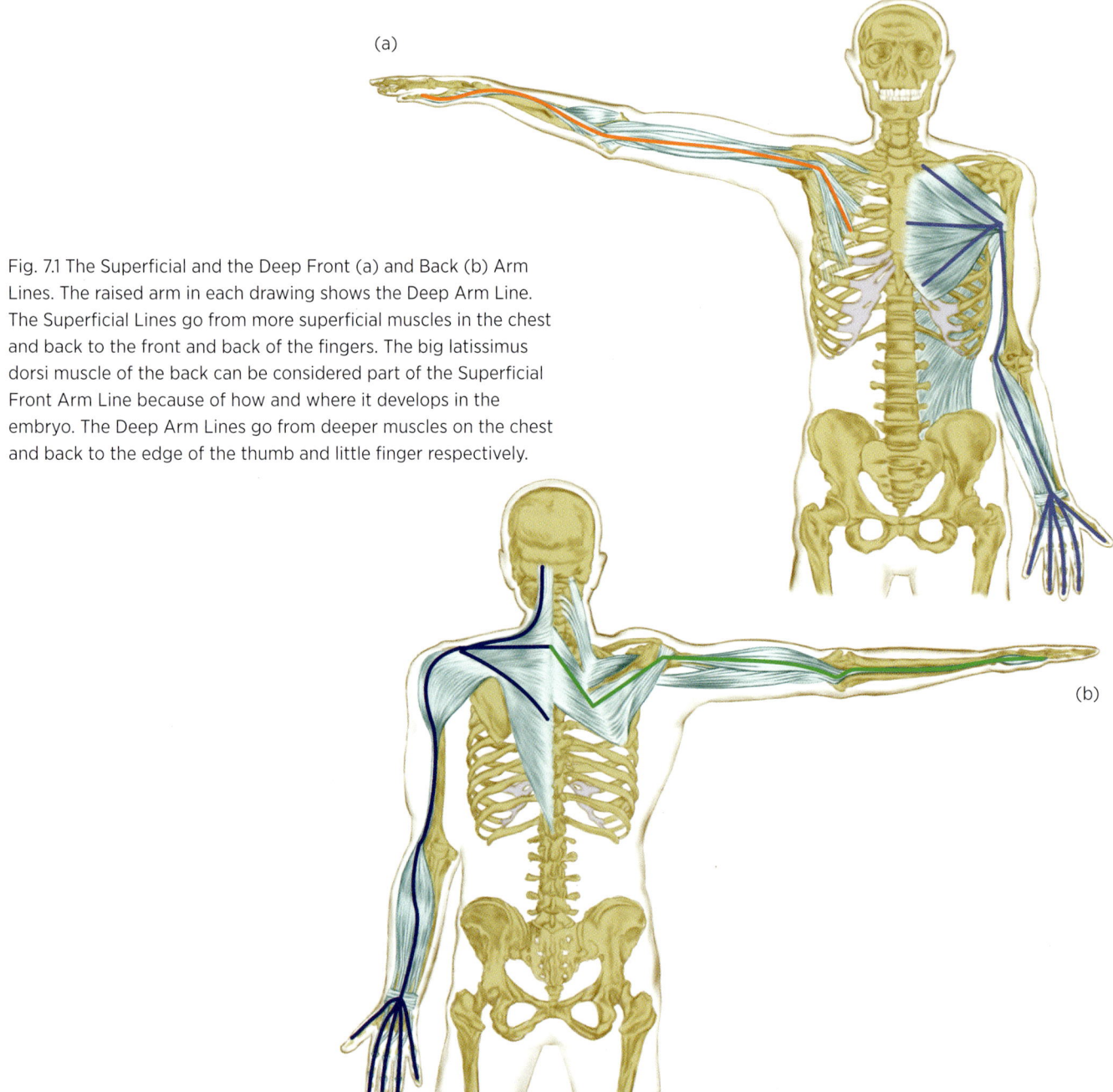

Fig. 7.1 The Superficial and the Deep Front (a) and Back (b) Arm Lines. The raised arm in each drawing shows the Deep Arm Line. The Superficial Lines go from more superficial muscles in the chest and back to the front and back of the fingers. The big latissimus dorsi muscle of the back can be considered part of the Superficial Front Arm Line because of how and where it develops in the embryo. The Deep Arm Lines go from deeper muscles on the chest and back to the edge of the thumb and little finger respectively.

We have already seen that they act as extensions of the Functional Lines, so that each arm is stabilised from the opposite pelvis and thigh. Since the shoulders rest on the ribs, disorganisation – or symmetry and stability – are passed into the Arm Lines (ALs) from the other lines. This means that there is much more to arm placement than the arms themselves. In practice, we all know this! By clarifying *how the arms attach to the torso,* we see how other lines influence the fate of the elbows, wrists and hands.

Essentially there are four ALs in each arm, connecting the upper torso to the outer side of the thumb, the outer side of the little finger, the palm, and the back of the hand. The Deep Front Arm Line (DFAL) connects the deeper muscles of the chest to the thumb, and the Deep Back Arm Line (DBAL) connects deep muscles of the back and neck (beneath the shoulder-blade) to the 'karate chop' side of the hand, and the little finger. The Superficial Front Arm Line (SFAL) connects the more superficial muscles in the chest to the palm and front of the fingertips, whilst the Superficial Back Arm Line (SBAL) connects the more superficial muscles of the back to the back of the hand and fingertips (see Figs. 7.1 and 7.2).

Fig. 7.2 The 'points in your pecs' where the Deep Front Arm Lines insert are also on the Functional Lines and Deep Front Line, as well on the boards of the intermediate stability system, part of the Lateral Lines.

The SFALs begin from the bony attachments at the top, middle and bottom of the large pectoral muscle of the upper chest, whilst the DFALs begin in the origin of the smaller (minor) pectoral muscle that lies beneath it. The minor muscle has 'fingers' inserting into your ribs on the line of the board, and it connects the Deep Front Arm Line (DFAL) to the intermediate stability system (see page 107). It also connects it to the Deep Front Line, which we will meet in Chapter 9.

The SBALs begin in the trapezius muscle. Together, the left and right trapezius create a diamond shape on your back; its mid-line extends along your spine from the bone at the back of your skull to your lowest rib, giving the line a wide sweep of attachments. The DBALs begin from the rhomboid muscle that lies beneath it. This is attached to the spinal processes of the bottom vertebra of your neck, and the first five spinous processes of the vertebrae in your ribcage.

Human shoulders and arms are designed for mobility, whilst legs require more stability. Thus arms have many more options than legs, and for us as riders, they have too many for our own good! They can hold things steady, bring things towards us, or push them away. If the hands grasp a steady object, they can also pull, push, or stabilise the body. The DFAL is primarily a stabilising line, connecting the thumb to the chest and controlling the angle of the hand. It determines if we ride keeping our 'thumbs on top'.

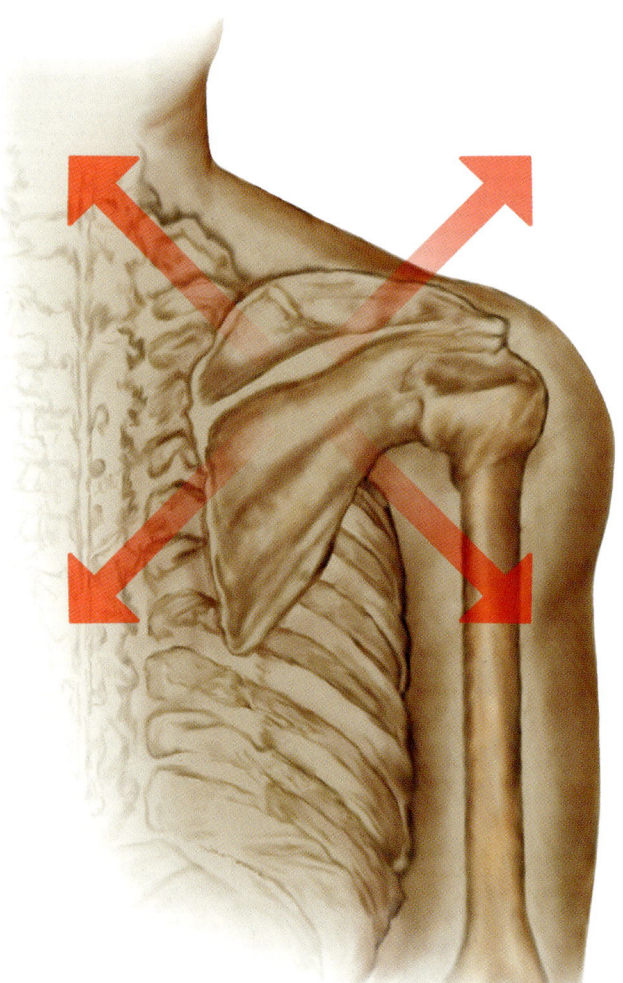

Fig. 7.3 The shoulder-blades are pulled in many directions, but we can simplify this by thinking of an 'X', the legs of which pull either up and in/down and out, or up and out/down and in.

Fig. 7.4 Pulling down your shoulders like this pulls the shoulder-blades down and in, and helps to mitigate the hours spent slouching over a computer.

Additionally, since our shoulder-blades ride freely on our back ribs, they can be pulled in a variety of directions by a variety of lines. The main four lines of pull form an 'X': one of its legs pulls them either down and out, or up and in. The other leg brings them down and in, or up and out (the latter from a short pectoralis minor, which pulls on the very front of the shoulder-blade to make your chest collapse and your shoulders round). Hunching over our computers activates the latter pattern – to find the antidote, put your arms behind your back, clasp your hands together, and pull them down to draw your shoulder-blades down and in (see Fig. 7.4).

As we have seen, the Deep Arm Lines (DALs) are attached to deeper muscles in the torso and (for the back arm lines) the neck – hence their name. These are the lines that stabilise the arms, and the difference in construction between them and the more superficial arm lines is highly significant for riders. The DALs (both front and back) have most of their muscles in the upper arms, whilst in the lower arms they are composed of a fibrous band that lies alongside the arm bones (the radius and ulna). The SALs have the opposite structure, with fibrous bands taking them from the shoulder to the elbow, and their more muscular part in the lower arm.

Misbehaving Arms and Hands

When the torso is working in the optimal way, it is easy to access the DALs, bending your elbows and pushing your hands out in front of you, with your thumbs on top – just as you have been taught. More often, we rely on the SALs, which are likely to create contortions in the wrists and elbows – in fact wrists allow so many aberrations that I often think we as riders would be better off without them! (see Fig. 7.5).

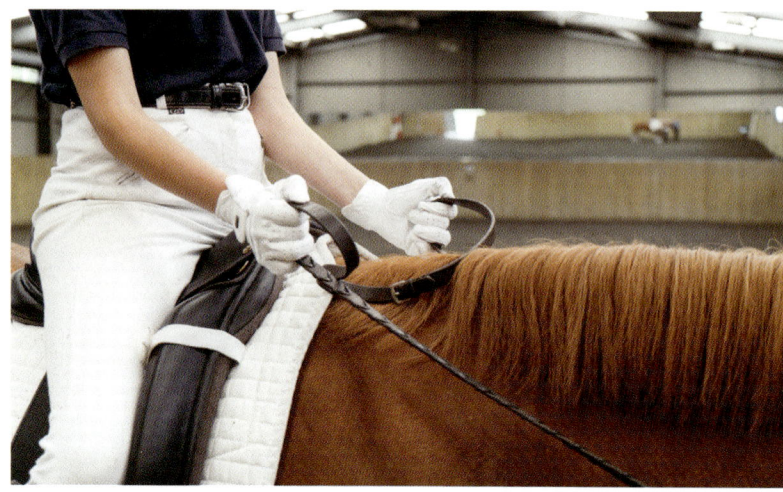

Fig. 7.5 Wrists! One or both wrists may suffer from these afflictions: (a) too in – which puts the fingers and knuckles too out.

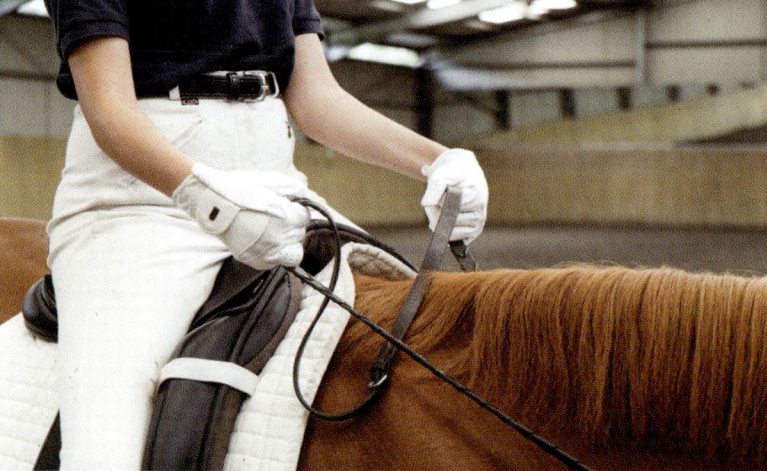

(b) Too out – which puts the fingers and knuckles too much in.

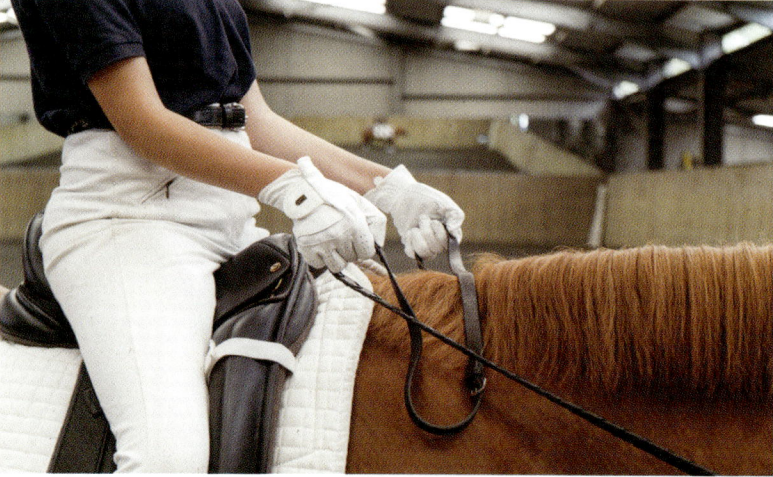

(c) Too up – which puts the fingers too down and aims the knuckles to the ground.

(d) Too down – which puts the fingers too up and aims the knuckles to the horse's eyes (this often goes with straight elbows).

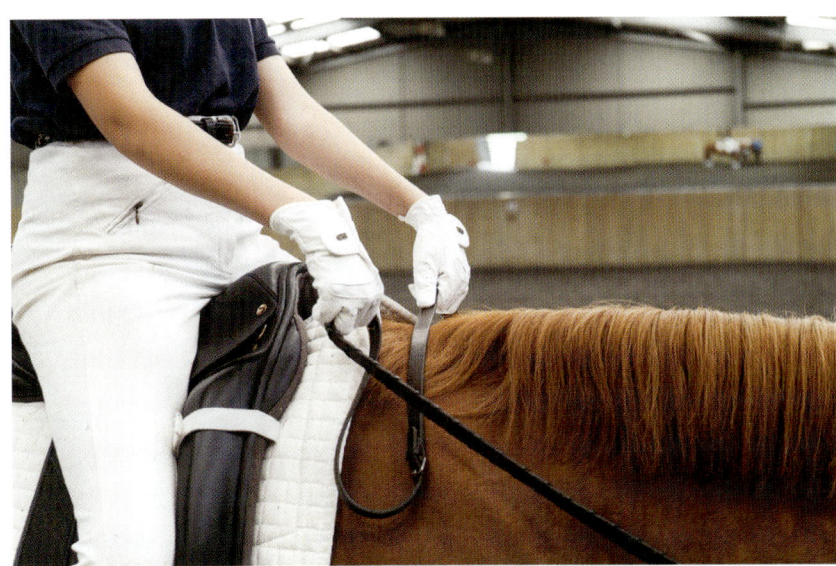

(e) 'Begging position' – which always goes with loose reins.

(f) Just right, with the back of the hand continuing the straight line of the arm, and the rein and lower arm making a straight line from the rider's elbow to the bit. If the rider's lower arm could elongate like Pinocchio's nose, the three fingers around the rein could wrap around the bit ring!

Organising the Wrists and Elbows

The closest I have been able to come to eliminating wrists is to invite people to ride for a few days wearing carpel tunnel supports. Without these, wrists can curl the hands towards each other, and/or make them go too much down, too much up or too much out. Most of these contortions are ways of taking up the slack in reins that are too long (see Fig. 7.5 previous page). Alternatively, the hands might go into 'begging position' with the fingernails down, which ensures that the reins will always be loopy. To the above contortions we can add straight arms, limp arms, floaty arms, and flapping elbows – and inevitably, very few riders have arms and hands that form a matching pair!

The following will help you organise your wrists and elbows:

1. If your arms look tense to an observer, have her identify where she thinks that tension lies. Does it primarily affect the elbows or wrists?
2. Think of your lower arms like Pinocchio's nose. If they could grow by magic, what direction would your hands go in? Would your knuckles aim towards the horse's eyes, the ground, or where else?
3. Arrange your wrists so that, if your arms did extend, you could wrap your first three fingers around the bit rings – without pulling on them, of course. (Most riders like this idea, dreaming of the control it would give them!) Imagining this keeps your knuckles pointing towards the bit rings, and will fix many of the aberrations to which wrists are prone (see Fig. 7.5 previous page).
4. If you keep turning your hands over so that your fingernails point down, use one hand as you are reading this to feel the inside of your opposite elbow. You will find a big bony knobble; keep that by your side, or even pressed into it. This is much more effective than attempting to fix the wrists by focusing on the wrists themselves. It will also fix elbows that flap, or drift out and up.
5. If your elbows are still not hanging under gravity, have your friend wobble them for you, until you can both feel that there is no resistance to this. Then tense them again, and feel the resistance this creates. Wobble them again until they just hang, and repeat this a few times. You may also benefit from pressing your elbows both in and out against a resistance provided by your friend. Which feels more familiar? (see Fig. 7.6).
6. You can adapt this idea to wrists, wobbling them and resisting the wobble, etc.

If you have one arm that will still not behave, ask yourself:

1. Is this the result of a steering drama? Are you unable to find a contact at the end of the outside rein? This is likely to make your lower arm and wrist 'droop'. Are you tempted to pull on the inside rein?
2. If one arm wobbles about like an 'alien arm', what is that arm attached to? Think of your thumb attaching, via the DFAL to the 'point in your pecs' on the line of the board. Similarly, think of your little finger and the DBAL inserting under your shoulder-blade, on the back of the board. Does this help? (see Fig. 7.2 page 135).

Fig. 7.6 When a friend holds your arm just above and below your elbow you can feel the difference between the tension that stops her from wobbling it, and letting go in a way that allows her to wobble it. Also press in and out against her resistance, until you can allow your elbows to just hang.

Even with these ideas, which are unusually effective, not many riders can just do what they're told and make an effective, lasting correction their hand position! Willpower and obedience (in other words, our conscious mind) will always fail us – it is no match for the task. It does not matter how many times you are told to 'Stop pulling', 'Shorten your reins up', 'Keep your hands still', or 'Take up a contact on the outside rein' – you will soon be disorganised into your favourite habit. You can never eradicate a symptom without uncovering the deeper cause.

Matching the Forces, Pushing the Hands

Our asymmetry is pivotal to our problems with arms and reins, but alongside it lies the fundamental, hidden demand of skilled riding, which is to match the forces of the horse's movement in each successive stride. Think of bouncing along on a space-hopper ball (a hippity-hop): the body has to be proactive in this hopping – if you just sat there and did nothing, you would fall off the back of it as it bounced away from you. When riding, many people act as if the reins were there to save them! This comes in the category of 'sad but true'.

Pulling on the reins – which seems like an arm problem – is actually a pelvis and back problem. Riders pull because they have not mastered the challenge of 'hopping along', matching the forces that the horse's movement exerts on their body. Only this enables them to push their hands forward – the traditional demonstration of 'an independent seat'. Either you push your big, heavy torso towards the hands (and 'hop' along) or you draw the light, mobile hands back towards the torso (whilst not 'hopping' effectively). This is comparable to the difference between just moving your hand and arm backwards, and attempting to push an immovable object across the floor! The latter is hugely demanding on muscle-power. Fig. 7.9(d) page 147 shows the rider pushing her hands forward against a resistance – practise this again and again – push against your steering wheel, your kitchen counters, and any surface you can find!

Furthermore, the odds are stacked against us, not only in the challenge of matching the forces, but also at the fundamental level of neurology. We manipulate the world with our hands: we use them to solve almost all of our problems, and we expect this to work on horses. The bias to use our hands is inbuilt, since a huge amount of our sensory and motor cortex are dedicated to the messages sent to and from our hands. This is shown pictorially in the homunculus of Fig. 7.7. Very little of the cortex is dedicated to feeling and moving the pelvis etc., but it is here that we need to become 'dextrous'.

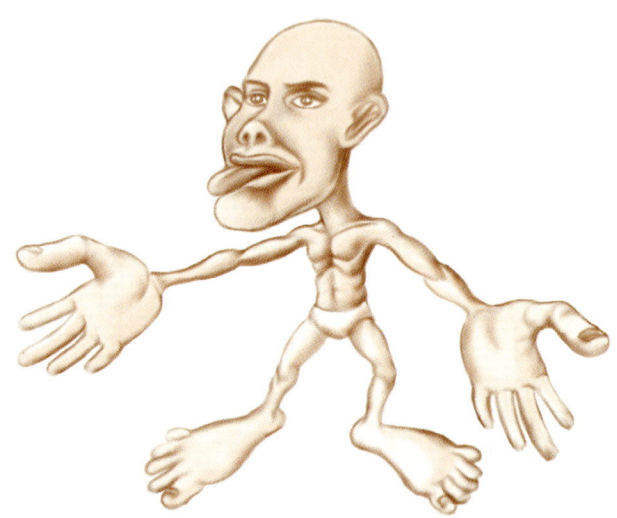

Fig. 7.7 The homunculus: this is how your sensory cortex maps your body, with much more 'real estate' dedicated to your hands, tongue, lips and mouth than to your pelvis and torso.

Remarkable as it seems, a significant percentage of riders live in total ignorance of their permanent pull on the reins! Whether they acknowledge this or not, these riders are nervous. To give the hand forward is – it would seem – to give up control, and this is a big 'ask' for many. I hope this book will help you discover (with less blood, toil, sweat and tears than it often takes) that control lies in the other lines of pull, and not the ALs.

1. When you ride, is the forward push of your abdominal 'bear down' stronger (see exercise page 62) or is your backward pull on the reins stronger? This is determined by whether you are 'with' your 'hippity-hop', matching the forces in each step that your horse takes, or are trying instead to stop him from bouncing away from you. You need to develop an awareness of when the pull becomes stronger – and a conscience about this too!
2. If you are not immediately successful, *keep searching for ways* to keep yourself on the correct balance point, perhaps by arching your low back a little to bring your weight on top of the horse's 'hippity-hop'. It is when you are toppling back and not matching the forces that you feel the need to pull (see Fig. 3.5 page 55). Be honest in your assessments, and realise that it is all down to physics!

3. Ask your coach or a friend to stand in front of your horse and hold the reins near the bit rings. Then have her ask you to 'take up a contact'. Does she sense your contact as a pull? If so, do you pull primarily with your elbows, or by drawing your hands down and/or back?
4. The contact needs to happen by virtue of where you place your hands, and the length of the reins.
5. Practise taking up a contact again and again until this is the case. Have your coach check that, when looking from the side, there is a straight line from your elbow to the horse's mouth, with your elbows slightly ahead of the side seams of your shirt (see Fig. 7.5f).
6. Check that you are holding the reins so that they cover the first bone in your fingers, and not the second bone (see Fig. 7.8a). Bend the first joint and keep the second joint straight so that the pads of your fingers (and not your fingernails) are against your palms. Keep the middle joint of your thumb up, so that your thumb makes a roof shape (see Fig. 7.8b).
7. Are you primarily holding the reins between your thumb and first finger, or between your ring finger and little finger? Or are you grasping them with all of your fingers, but no thumb pressure? Teach yourself to hold them between the thumb and first finger.

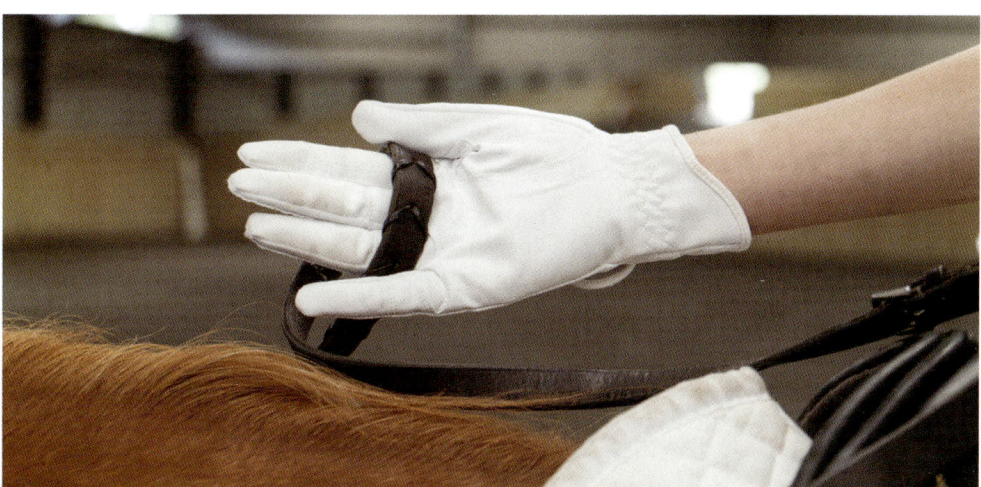

Fig. 7.8 (a) Hold the reins so that they cover the first bone of your fingers, with the second joint straight and your thumb in a roof shape.

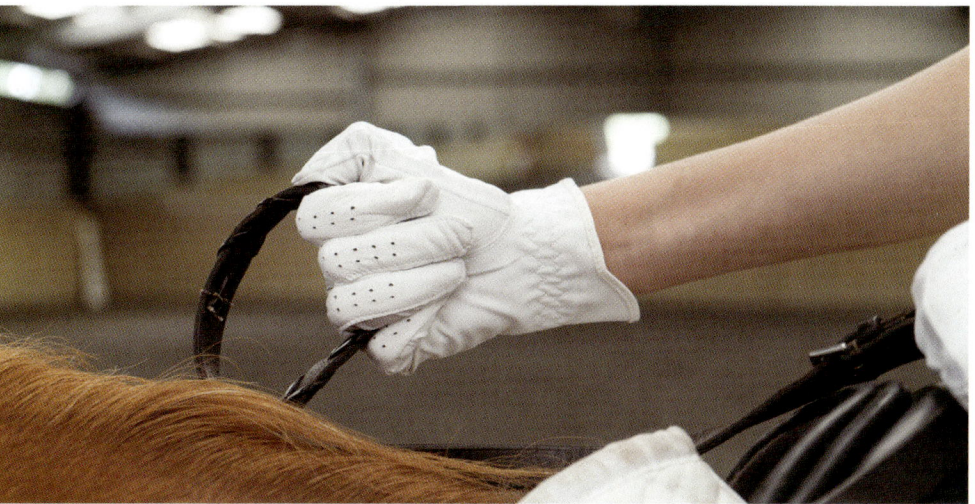

(b) The second joint in your fingers should be straight, so that your finger pads are against your palms, not your fingernails. Your thumbs should make a roof shape.

Passive Resistance

To understand how the contact can become stronger when necessary without becoming a pull, you need to be able to do a 'passive resistance'.

1. When you are sitting in 'neutral spine', ask your coach to hold the reins near the bit and to gradually make a pull on them.
2. As your coach pulls, press harder with your thumb on your first finger, and activate your armpit tendons (see Fig. 3.12 page 69 and Fig. 6.2 page 124). Think of your trapezius muscles, which form a diamond shape with its point at your waist, and pull your shoulder-blades down towards that point. You then involve both BALs.
3. When you do this well, your arms act like side-reins, which never actually pull (ever danced the 'jitterbug'?). Instead, your horse has to sort out his movement in relation to them. Together with your coach, clarify the difference in feel between pulling back, and making this 'passive resistance'.
4. Your coach can test for this by suddenly letting go of the reins. If you fall backwards you were pulling! If you were making a good 'passive resistance' you will not move.

Halts and downward transitions will never work well if you cannot do a passive resistance, but inevitably, there is more to it!

1. The passive resistance is the last part of your halt aid. Learning to halt well (and also to half-halt) involves *stopping yourself from doing the many possible wrong things*, so that the right things can emerge. Sadly, many riders have paid a lot of money to learn the wrong things!
2. To ride a good halt, most riders need their reins shorter than they think they will.
3. It is important not to invoke the water-ski/motor boat dynamic. So you must remain vertical, resisting a push on your upper chest (see Fig. 3.7 page 61). Do not lean back, or hollow your back or round your back: remain in neutral spine throughout.
4. You will need visual feedback to confirm that you are indeed doing this. Most riders drop back behind vertical when riding halt, and feel as if they are leaning forwards when they actually remain neutral.
5. Do not grow tall and/or push down into your stirrups. Snug your thighs in more firmly instead. If you are a hollow-backed rider, keep your chest and knees close together.
6. Do not allow yourself to go floppy, or allow the horse to slow down his legs and 'wind down' into the halt as if he were a clockwork toy. He has to go 'walk, walk, stop'(and you both need to become reliably good at this before you practice halting from trot). A good halt is reminiscent of a knife going through butter.
7. Stop your seat bone movement, and think of stopping the horse as you would stop a trampoline. Bear down, and breathe out. Make the sound 'Psssht'.
8. Use a passive resistance on the reins by simply pressing harder with your thumbs. This will easily morph into a pull if your reins are too long and you

draw your hands back to take up the slack. You will also end up pulling if you push down into your stirrups, or distort your vertical, box-like torso in any way.

9. If you have to hold the passive resistance for a long time, do so. If it really does not work, diagnose yourself: did you suck in your stomach in, grow tall, push in your stirrups, lean back, or invent another aberration? A lot of these aberrations disappear if you bear down and maintain neutral spine!

Short Reins and 'Arm Cuffs'

As Charlotte Dujardin once said in a dressage symposium, 'Short reins win medals!' But left to themselves, very few riders keep their reins short enough. It is ironic that most riders start to pull when they bring their hands backwards to take up the slack in reins that are already *too long*. Short reins rarely lead to pulling (except perhaps in the case of the very nervous novice). Meanwhile, trainers find themselves saying 'Shorten your reins up.' again and again. The correction simply does not last. Despite all their best intentions, riders soon find themselves pulling back on reins that are too long.

In most riders, this 'arm' problem happens because the forces of the horse's movement create a disconnection within the rider's torso-box that, in effect, makes the *back of it fall away from its front*. The BALs go backwards as well, and the reins get longer and longer. Instead of being held between the thumb and first finger, they often begin to be held between the little finger and ring finger. The latter is one of the sneakiest ways in which riders morph into pulling.

When the back of your torso-box is firmly enough attached to the front of the box it becomes much easier to push your hands away from you, keeping your reins short enough. Realise too that *only then do you stand a reasonable chance of connecting the back of your horse to the front of your horse*. This is never going to happen if you do not have enough strength to connect your own back to your own front!

To help you do this, we are going to invent 'arm cuffs', shaped like handcuffs. A pair of arm cuffs on each side keeps your upper arms linked to your lats. One of the cuffs goes around your upper arm about one-third of the way down it and the other is body-pierced around the bulk of the latissimus muscle below the back of the armpit. (Sorry!) Imagine one link of chain (or a short piece of very strong elastic) between the cuffs.

1. To keep the cuff on your arm pulling on the cuff around your lats, you need to have your elbow slightly ahead of the side seam of your shirt, otherwise the link between them goes slack. This positioning of the elbow is crucial.
2. If you are a rider whose elbows and hands pull backwards, maintaining a constant pull on the cuff around your lats will ensure that your hands and elbows push forward instead.
3. If you are a rider whose elbows float forward, up and away from your torso so that your reins are always loopy, this idea will also help you maintain a contact. Your elbows can only move a little forward before they meet the (imaginary) resistance of the cuff.
4. If you suffer from the latter affliction, also work on firming up the Superficial

Front Line. Keep thinking of resisting a push on your upper chest, and learn to firm up the wall made by your abdominal muscles. Almost certainly, it too is 'floaty'! Try some of the sounds from page 65.

5. If you still tend to pull, and the arm cuffs do not solve your problem, you need to delve more deeply into anatomy to find the strength you need. Begin by re-finding the tendons at the front and back of the armpit. With both hands in a riding position, pull down both tendons as you pull your elbows down. Feel the back and front of your armpits firm up.

6. Imagine the tendon at the back of the armpit being pulled towards the tendon at the front of it, as if tightly tied with a shoelace. You could even imagine them both being pulled towards your horse's ears (without leaning forward).

7. Ask your coach to give you resistances under your elbows and then on the front of your upper arm, so you can push into her arm cuffs. Especially if you tend to drop your wrists or hands, add the resistance of your coach's straight arm above both wrists and then have her offer you a resistance, in front of your knuckles (see Fig. 7.9).

Fig. 7.9(a) The tendons at the front and back of the armpit firm up when you push the elbows down into a resistance.

(b) The coach offers a resistance on the upper arm that mimics the arm cuffs.

(c) It helps many riders to resist the wrists upwards. This a really good fix if you straighten your elbows.

(d) Everyone benefits from pushing their hands forward against a resistance.

(e) The diagonal line connects a point just under and above the tip of the shoulder-blades with the 'point in the pecs'.

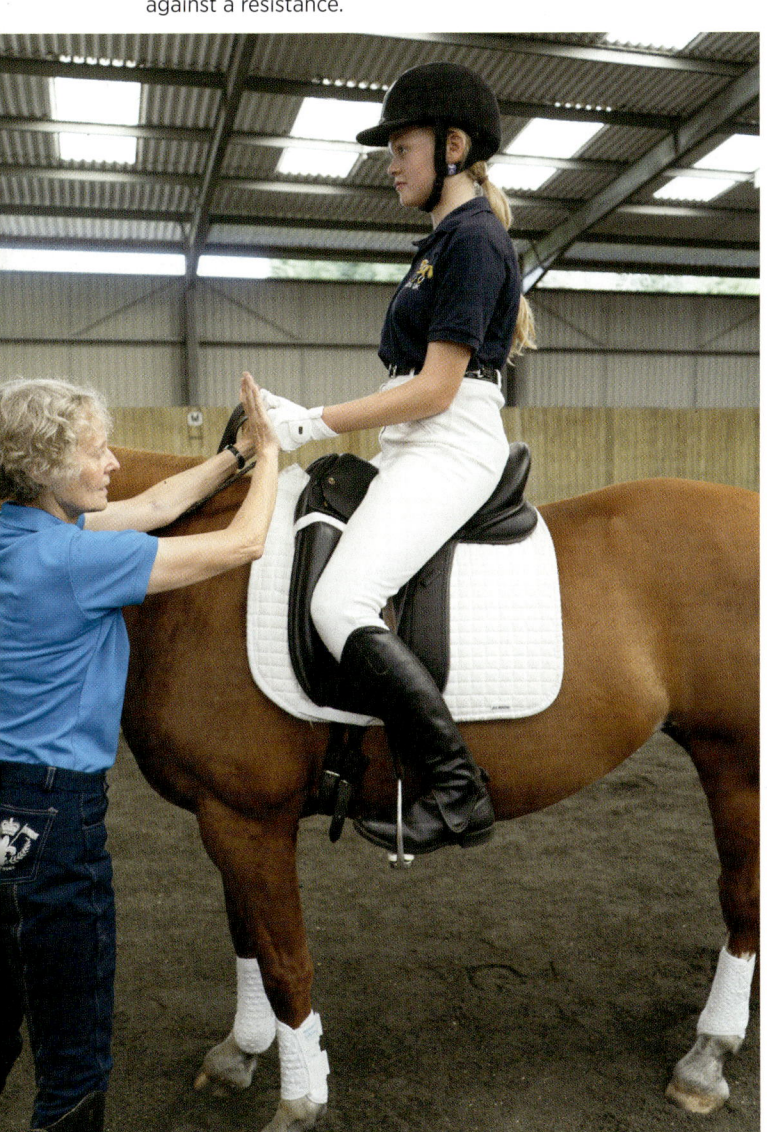

If this still fails, the problem is that the points of your shoulder-blades (in particular) are falling back away from your torso, and taking your arms with them. In this case:

1. Imagine an attachment within you that pulls a point under and just above the bottom of the shoulder-blade (see Figs. 7.9(e) and 7.10) towards the 'point in your pecs' where the DFAL inserts (see Fig. 7.2 page 135). Think of your thumb being connected into this point in front, and your little finger connected to this point on your back, via the DBAL .
2. Thus, you are pulling the insertion of DBAL towards the insertion of the DFAL. Pull down both of your armpit tendons and push your hands forward. For some people, this is the key idea that suddenly makes it possible to access both DALs and push the hands forward.

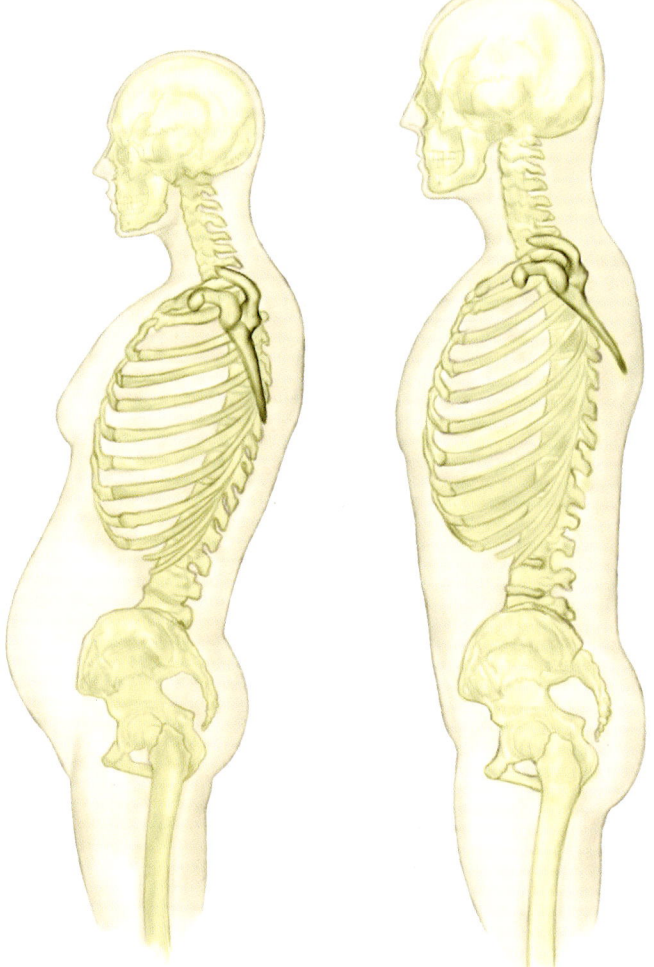

Fig. 7.10 If the ribcage is tilted backwards the shoulder-blades may appear flat, but they are in fact tilted, pulled forward by a short pectoralis minor. Thinking of the tips of them being pulled towards the points in your pecs can make it much easier to push the hand forward.

The Contact Scale

Your aim is that it is you, not the horse, who determines how much rein contact you have on a scale from 0 to 10. A number of exercises follow, because between the rider pulling, the horse pulling, the horse putting a loop in the rein etc., there are a number of problems that all riders face.

1. Define this scale with your coach holding the reins near the bit, so you can communicate without confusion. (Without this, how can one of you know what the other ones means by a 'strong' or 'light' contact?)
2. You do not always want a 1, especially with horses who would like to offer a very light contact. You certainly do not want a 10, but you need to be more able than you might think to navigate this scale, *without the contact becoming a pull.*
3. Remember how the rein must pass through your hand, held primarily between your thumb and first finger (see Fig. 7.8 page 143).
4. Your job is to keep the bit stable in the corners of the horse's mouth, so you can feel it there, and the horse can feel your consistent contact. It can help to think of holding the hand of a child, or to sense that contact in your elbows rather than your hands.
5. If the contact needs to become stronger, the rule is still that the push of your abdominal 'bear down' must remain stronger than your hold on the rein. Your core has to strengthen before your arm can handle a stronger contact without pulling.

All of us need to discover what lies between the all-too-common options of loopy reins and pulling. Riders who are addicted to having loopy reins often need convincing that is it not cruel to 'take up a contact'. I ask them to imagine that they have no bridle on their horse, but only the bit and reins. It would be horribly cruel to allow the bit to bang against the horse's teeth or fall out of his mouth!

I have also convinced riders of the value of a steady contact by play-acting a poor cell phone connection. 'Hello, are you there? You're breaking up, I can't hear you? Hello? That's better, I can hear you now.... Oh no, you've gone again...' We all know how frustrating this can be!

When all else fails, I have been known to stand in the middle of the rider's circle saying 'loop', 'loop, loop' (they often come in pairs) whenever a loop appears in the rein. This annoys people so much that it becomes a short-cut to much greater motivation!

Shortening the Reins

Shortening the reins well is an art form, especially given that so many horses have good reason to be 'shortening-the-reins phobic'! The problem is often that riders take too much of their awareness *away from everything else* when they shorten the reins.

1. First, realise that a shorter rein does not necessarily mean a stronger contact. It means having your arm and hand more out in front of you, with your elbows slightly ahead of the side seam of your shirt (and pulling on the imaginary arm cuffs).
2. As you shorten the reins, keep at least 80 per cent of your attention on your body – bear down, thighs on, feet light, etc.
3. Shorten your reins with no more than 20 per cent of your attention.

4. Practise shortening the reins at walk: the horse must not know that you are doing it. He will realise if your body changes, and/or if the reins become either loopy or tight in the process.
5. Put one rein under the other thumb as you slide your fingers along the rein. Reins with leather stops can help you be more aware of the length of your reins.
6. Double-check: are you now pulling? Or do you just have a shorter rein with your hand and arm more out in front of you? Keep pushing your hands forward against an imaginary resistance, as you hop along on your 'hippity- hop'.

When the Horse Puts a Loop in the Reins

Your arm cuffs keep your hands and elbows pushing forward, but since your horse might make his neck shorter and put a loop in the reins, you also need a strategy for taking up the slack in the reins without pulling back. On horses with an unsteady head-carriage, this is tricky, but it is an essential step towards changing their pattern.

1. Ask your coach or a friend to stand in front of your horse and take hold of the reins near the bit rings so that she can play the role of the horse. The reins need to be long enough for your helper to have a loop between her hands and the horse's mouth, whilst both of you have contact with each other (without pulling!).
2. Your helper then moves her hands towards you, putting a loop into the reins.
3. How do you react? Do your hands come back towards your body, or do you have a different strategy to take up the slack? Do the reins remain loopy?
4. Taking out the slack in the reins is best done by bringing your hands out sideways, as you would if you played an accordion. Your elbows stay in place, with your fingernails, if anything, facing up slightly as your hands open. It is important that your knuckles do not face down, and your elbows do not straighten (see Fig. 7.11).
5. Can you respond to the potential loop in the reins so promptly that the loop never appears, but so smoothly that the bit just stays in the corners of the horse's mouth? It usually takes quite a few repetitions for a rider to become able to recognise the *tiny beginnings* of slack coming into the rein, and to widen her hands in order to takes up that slack smoothly before it ever materialises.
6. Practise this repeatedly, and then also with your eyes closed so that you only have your felt-sense to guide you.
7. When you ride, be ready for the moment when you *might have to* widen your hands to take slack out of the rein.
8. The ideal width apart for the hands is determined by the idea of a straight line made by the lower arms and reins when viewed from above (in addition to the more obvious straight line from the rider's elbow to the horse's mouth, which an observer would see from the side) (see Figs. 7.11(a), 7.5(f) page 139).
9. If you find yourself continually riding with your hands wider apart than the ideal shown in Fig. 7.11(b), you need to shorten up your reins and begin again.

Fig. 7.11 Widening the hands apart to take up the slack when the horse scrunches his neck backwards and puts a loop in the reins. (a) Ideally, when seen from above, the lower arms and reins make a long, thin triangle.

(b) The rider's hands widen apart to take up the slack that I (as the horse) have put into the reins. The fingernails must point slightly up and out, not down.

When the Horse Offers to Lengthen His Neck

It is less tricky to advance your hands and elbows slightly when the horse offers to lengthen his neck, or to slip the reins a little. It is likely, however, that his neck will shorten again, requiring you to either widen your hands apart, or shorten your reins. Elongating his neck when he prefers to 'push back' at you (see page 86) draws on the skills of Chapters 3 and 4. It is really about the influence of your abdominal bear down (see page 62), which can create a forward force that overcomes the backward force of his 'push back'. He then reaches his neck away from you, elongating his Superficial Back Line. Concurrently his back lifts, as he becomes easier to ride, and more beautiful to watch.[1]

When the Horse Pulls

Groundwork can be the best way to show your horse the meaning of the legs as 'go' and the reins as 'stop', and also to retrain the 'go' and 'stop' responses of horses who either do not understand, or do not take the rider seriously. It is sad but true that a remarkable percentage of horses do not have a conscious understanding of this![2]

Whenever the horse wants to pull on you, we could say that you have a problem with 'stop'. Pulling against his pull is never the best answer, given that he is so much bigger than you, and can quite happily keep pulling for much, much longer than you can!

1. Think of your arm(s) attaching, via the DFAL, to the 'point in your pecs' where the line inserts (see Figs. 7.1, 7.2 pages 134-135). As you attach your arm here, it becomes easier not to meet the horse's strength with strength that comes from your arm muscles and the sides of your torso-box. This tactic can work like magic to lighten a strong contact, especially if the horse pulls on only one rein.
2. If you and the horse are forming a counterbalance – like two people holding hands and leaning away from each other – you need to *stop playing this game*! If

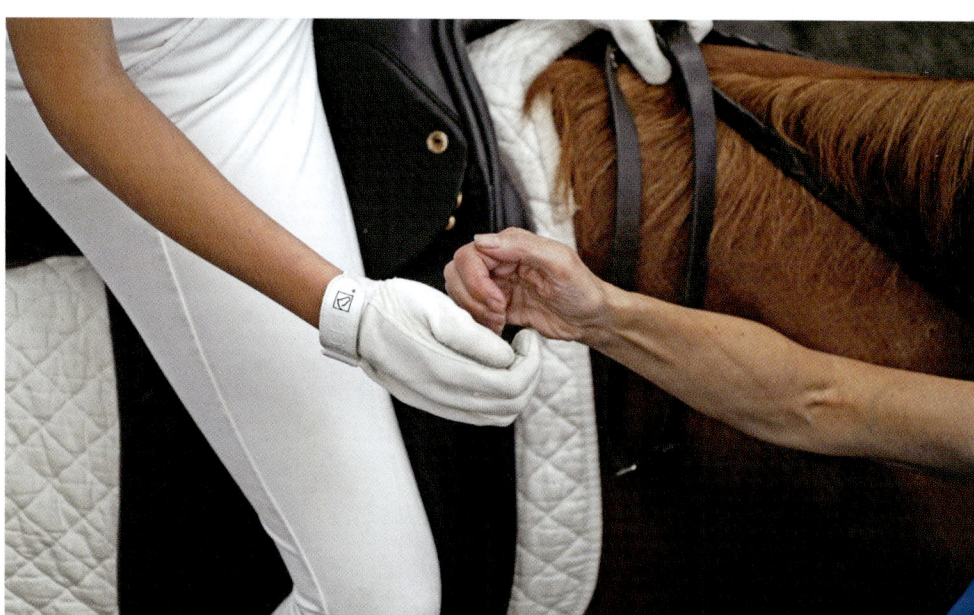

Fig. 7.12(a) Rider and coach hold hands by curling their fingers around each other.

When the coach leans back and pulls, most riders will lean back and also pull, as in (b).

you refuse to hold the horse up, he will have no choice but to hold himself up. This requires you to be in balance in your own right, independently of him. This forces him to take responsibility as well.

3. Practise the following, ideally in halt, with you and your coach holding hands as shown in Fig. 7.12. Ask your coach to lean back against you and make a pull – how would you get out of this?

4. There is one option that few riders think of: *give your hand forward quickly,* and hold it forward for about three seconds. Expect your coach to have to take a step back to put her feet under her centre of gravity! If the coach can adapt to the movement of your arm, you did it too slowly. Be careful not to lean forward or round your back, etc. as you give your hand forward – your torso-box must remain engaged.

But as in (c), they need to give the hand forward and refuse to play the horse's 'game'.

In (d), the coach, like the horse in the same situation, has to take responsibility for holding herself up – as does the rider.

5. Bring your hand slowly back into place. Your coach is the judge of whether you re-created a pull back as you did so. It may take a number of repetitions before you can put your hand in place with no pull. I have been known to spend twenty minutes working with a rider before she could do this well!
6. Repeat this whilst walking your horse, giving your inside hand towards the horse's mouth (not his eyes or the ground). It has to move forward quickly, and stay forward for the time that it takes for both of his forelegs to step. Bring your hand back into place slowly and carefully.
7. How many strides does this 'buy' you before your horse invites you to pull again?
8. Repeat giving your hand forward when this happens. If your horse curves his neck to the opposite side when you give your hand forward, you have a significant steering problem and need to get much more 'stuffing' in the side his neck curves towards, filling him out into a longer Lateral Line (see page 115).
9. Try giving your hand forward in trot, and if necessary in canter.
10. Realise that you have to remain vertical throughout (or leaning forward slightly if you are in rising trot) with the Superficial Front and Back Lines remaining stable.

Visualising the Reins as Rods

When you are riding, it should ideally feel as if the reins are solid rods that push the horse's nose away from you. But very few riders or trainers realise just how much torso and arm stability underlies this skill, along with robust, well-balanced Superficial Front and Back Lines that keep you stacked up and matching the forces of the horse's movement. With help from the intermediate stability system, and well-balanced Lateral, Functional Lines, Spiral Lines, and Deep Front Line, riding becomes like pushing a wheelbarrow, with no pulling on either rein. If you are thinking 'dream on', your salvation will come in the next two chapters!

The 'Arm Lines' in the Horse

The Danish researchers Vibeke Sødring Elrønd and Rikke Schultz[3] found two 'Arm Lines' (Front Limb Lines –FLLs) in the horse, lying on the front and back of the limb, and drawing it forward – the Front Limb Protraction Line(FLPL) – and back, the Front Limb Retraction Line (FLRL). The human ALs do not map very well onto either our (hind) legs or the horse's forelegs; but broadly speaking, his FLLs (see Fig. 7.13) are the correlates of the SFAL and SBAL in the human, since the horse walks, in effect, on the fingernail of his middle finger. The other toes were lost in evolutionary time, so the Deep Arm Lines in the horse are much reduced, but still present in the upper part of the limb. (They move the limb toward and away from the mid-line, and are the subject of current research).

The horse does not have collarbones, so his shoulder-blades are not attached to the rest of his skeleton. This allows them to glide in pendulum-like motions over the underlying tissue and ribs. Whilst human ALs determine how we reach, push

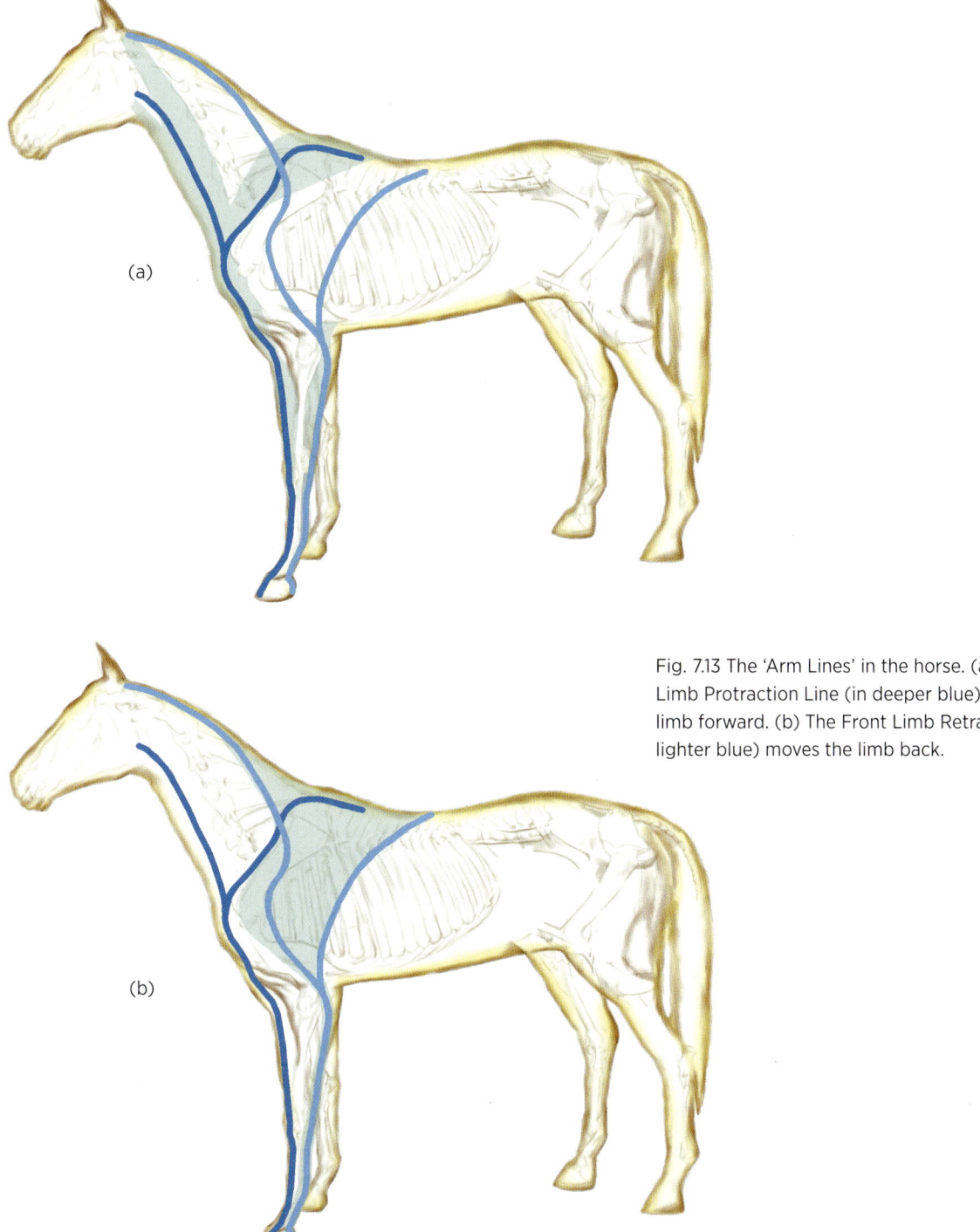

Fig. 7.13 The 'Arm Lines' in the horse. (a) The Front Limb Protraction Line (in deeper blue) moves the limb forward. (b) The Front Limb Retraction Line (in lighter blue) moves the limb back.

and pull, the horse's FLLs have instead to handle much of his weight (much more than the hind legs) – so the forelegs have a big influence on how he stands and steps. Modern breeding of sport horses has yielded forelimbs with a much increased range of movement, and Spanish walk (commonly trained in the Iberian tradition) encourages horses to maximise this.

Some horses love to do cat stretches, elongating the Front Limb Retraction Lines. They tend to have the hollow of the 'mantrap' (see page 86) just behind shoulder-blades, and the rider easily gets sucked into this. Conversely, when a horse piaffes as if on a pedestal, the Protraction Lines are elongated, and his forefeet come back under his mass, presumably in a tactic designed to take some of the load off his hind legs.

Professor Emerita Hilary Clayton[4] argues that the horse's forelegs are designed to produce more braking force, whilst hind legs provide more propulsion. She has demonstrated in her laboratory that, in collection, horses use their forelegs (and thus their FLLs) to push their weight back towards their hind end. Riders with good biomechanics can use their own body to assist this (pages 86 and 132).

Laterality

The DALs (which connect to our core) invite us to consider laterality in both humans and horses. This is not limited to our dominant hand and leg; our dominant eye and ear determine much about our personality and learning style as well as our asymmetry.[5]

Horses tend to peruse their environment through one eye at a time and, given that their eyes are on the sides of their head, looking ahead through a preferred eye requires them to elongate one side of their body. Some authors have considered this a significant factor in the horse's asymmetry.

Some French research has shown that horses are more spooky from their left eye (which sends input primarily to their right brain hemisphere).[6] Also, research has suggested that they listen to a friendly, known whinny primarily through their right ear, and to an unknown whinny through their left ear.[7]

We humans have one hand and one leg that we are more dextrous with, and one that we use more for support. So while one limb specialises in mobility, the other specialises in stability. (Which would you use to kick a football?) If the horse naturally bears more weight on one foreleg, perhaps we should call this his stabilising leg: he would halt first with that leg, and bring the other one up to it. He would then step off with that second (more dextrous) leg, using the stabilising leg for support.

To be consistent with the terminology we use in people, *his dominant leg would be the more dextrous one*. However, some authors have labelled the stabilising leg as dominant, and the ideas and terminology currently used by different authors are inconsistent and confusing.

Grazing Position and its Effects on Asymmetry

When foals graze, they have to spread their forelegs to reach the ground, as do many adult horses, who often show a preference for one foot being forward when they lower their head. The forefeet then develop differently, with a lower heel on the leg that is placed forward, and a higher heel on the leg that is placed back. More weight is placed towards the toe of this back hoof, and the heel has increased blood flow, which causes it to grow. The opposite is true of the hoof that is forward – more weight is placed towards the heels, which crushes them and causes the toe to grow longer.

The tissues of the entire FLRL of the leg that is held back are consistently shortened – this effect is not limited to the horse's flexor tendons. Meanwhile, the line at the front of this limb is relatively longer. The opposite is true of the forward limb, with its low heel being part of a longer Retraction Line. The shoulder of this forward, flatter foot tends to bulge out, and is held more forward. The shoulder of the more upright, held-back foot is flatter to the ribs and further back. To see any such asymmetry in your own horse, stand him as square as you can, and look at his shoulders from high up and behind him (see Fig. 7.14).

Carried on long enough, this sets up a pattern that affects the symmetry of the horse's gait. However, in her doctoral studies at Kassel University, Sandra Kuhnke has found that this preference does not determine the ridden horse's asymmetry, suggesting that in many cases the rider's asymmetry may be the dominant factor. Her study compared rein tensions, but remember that handedness may not be the issue; the arm that pulls more will belong to the side of the body that tends to be fall back (as the heel goes too forward). It is less able to transmit force from back to front and 'keep up with' the horse.[8]

Dr Andrew McLean, a founding member of the International Society for Equitation Science (ISES), has observed that one diagonal pair of legs tend to accelerate in the swing phase, i.e. whilst in the air.[9] He believes that 75 per cent of horses appear to be right foreleg runners (in the swing phase) and he observes that the right fore and left leg hind then do not shorten or slow as well as the other pair. (This involves the Functional Lines too.) In piaffe, this diagonal pair creeps forward in the swing phase whilst the other is lifted more vertically. I suspect that the horse who falls out onto that running foreleg when ridden on a circle will accelerate as he does so, whilst the horse who falls out onto the other foreleg is more likely to slow down.

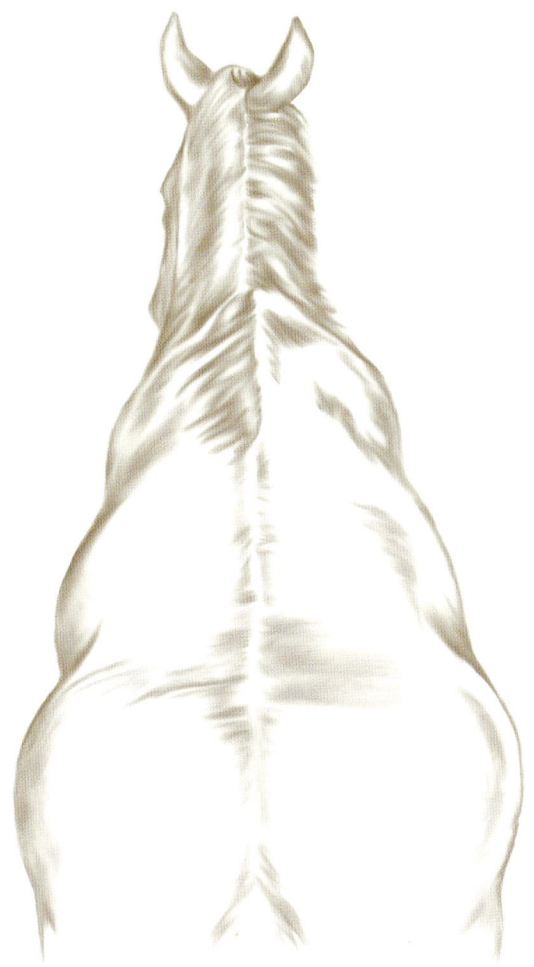

Fig. 7.14 Uneven shoulders are best seen from high up and behind. This horse will not want to stand square: his right foreleg has the lower heel and placing it forward helps him balance his body by shortening the distance between the shoulder and the ground. The mane falls to the lower, left, side, which has the higher heel. Placing the leg back helps him balance his body by lengthening the distance between the shoulder and the ground.

Minimising Asymmetry

The more we breed horses with long legs and small heads, the more commonly hoof, forelimb and shoulder asymmetry become a significant issue. Influencing this takes a village – your farrier, saddler, and equine bodyworker all have a part to play, although there is controversy about their most helpful interventions. To complicate the issue, there will be knock-on effects in the symmetry of horse's hindquarters via the Functional and Spiral Lines (see next chapter for the latter).

Your biomechanics coach has the least controversial role, her aim being to keep you (and the saddle) level enough for you to resist becoming disorganised by the horse's asymmetry, otherwise you will perpetuate it. Meanwhile, I hope that your team can maximise the horse's chances of being able to do what you are asking him to do, and that are you are skilled and disciplined enough to work your horse appropriately for each stage in his rehabilitation.

There will always be minor differences in horse's left and right Arm Lines, and mostly they are not significant. I have seen remarkable changes in horses with extremely asymmetrical feet and shoulders – they have become sound and functional even when the difference had previously led to lameness. That said, you and your veterinarian should look particularly carefully at the forefeet and shoulders of any horse you are tempted to buy.

CHAPTER 7 NOTES

1. Wanless, Mary, *Ride with your Mind Essentials*, Kenilworth Press (2001)
   Wanless, Mary *Ride with your Mind Clinic*, Kenilworth Press (2008)

2. McLean, Andrew www.hosemagasine.com/thm/..../principles-of-horse-training-with-andrew-mclean www.esi-education.com

3. Sødring Elbrønd, Vibeke and Schultz, Mark, Rikke, 'Myofascia – the unexplored tissue: myofascial kinetic lines in horses, a model for describing locomotion using comparative dissection studies derived from human lines', *Medical Research Archives* (2015) Issue 3

4. Clayton, H.M., H C Schamhardt and Hobbs, S.J., 'Ground reaction forces of elite dressage Horses in collected trot and passage', *The Veterinary Journal* (2017) 211 pp.30–3

5. Hannaford, Carla, *The Dominance Factor: How Knowing Your Dominant Eye Ear Brain Hand and Foot Can Improve your Learning*, 2nd edn, Edition, Anglo American Publishing (2011)

6. De Boyer des Roches, A., Richard-Yris, A.M., Henry S.,and Hausberger, M., 'Laterality and emotions: visual laterality in the domestic horse (Equus caballus) differs with objects' emotional value', *Physiology & Behavior* (2008), 94: pp.487–490

7. Basile, M., Boivin, S., Boutin, A. et al., 'Socially dependent auditory laterality in domestic horses (Equus caballus)' *Animal Cognition* (2009) 12:611 3 and Christa Lesté-Lasserre, MA, 'Listen up: horses demonstrate auditory laterality, (2010) www.thehorse.com/articles/24881/listen-up-horses-demonstrate-auditory-laterality

8. Kuhnke, Sandra, U. and von Borstel, König , 'A comparison of methods to determine equine laterality in thoroughbreds', *Journal of Veterinary Behaviour Applications and Research* (Sept. 2016) and 'A comparison of rein tension with methods to determine equine laterality', *Journal of Veterinary Behaviour Clinical Applications and Research* (Sept. 2016)

9. McLean, Andrew, 'Thinking about horses, Part 2', www.horsemagasine.com/htm/2015/01/andrew-mclean-thinking-about-horses/

PART 4 | TWISTS, TURNS AND THE REAL DEAL OF THE CORE

CHAPTER 8

THE SPIRAL LINES

An Introduction to the Spiral Lines

On first look, the left and right Spiral Lines (SPLs) are the most daunting of all the lines to follow, as they cross the mid-line from right to left (and vice versa) three times. Many of the muscles in the SPLs also participate in the lines we have already met, which simplifies our learning. It also means that the SPLs can add to the lateral stability provided by those lines. Unfortunately, it also means that dysfunction in the SPLs can compromise their functioning too.

The SPLs wrap around the core to form a double spiral, and thus they contribute to postural rotations and twists. However, firm lines hold the torso in place, and since they also run like stirrups under the arches of the feet, they stabilise the lower legs and feet. This increases your ability to feel and control them. The SPLs make a long and complex journey; so take a deep breath, and follow along with me as you look at Fig. 8.1.

Each SPL starts on the back of the skull behind your ear, sweeping with broad sheets of muscle on a diagonal path, crossing the spine over the lower two vertebrae of the neck, and the first five of the upper back. It then continues to touch the opposite shoulder-blade, and then pass *under* it to the side of your ribs. From the ribs, the SPL wraps diagonally across the belly to the point of the opposite hip. Having crossed the mid-line in the back and then the front, it has returned to the side from which it started, and will now wrap around the leg on this same side (see Fig. 8.1).

From the front of the hip bone, the line drops down the front/outside of the thigh and knee. To feel a crucial muscle in this line, put your thumb just below and outside the point on the front of your hip (the anterior superior iliac spine, or ASIS). Flex your hip and turn your knee in to feel this muscle fire. This tensor fasciae latae muscle is a key muscle for riders, and you will often feel a combination of tickle and pain when you press your thumb into it. Slide your hand down your thigh to the outside of your knee to feel the fascial band that connects to it. The iliotibial tract is the much-maligned target of a lot of 'fascial rolling'.

The SPL then crosses over the front of the shin to the inside of the foot. It forms a sling under the arch of the foot, named 'the stirrup' by the old German anatomists. However, it lies further back on the foot than we place the stirrup when riding – think of the loop that might go under your foot on a pair of slacks.

The loop continues to the outside of the arch of the foot, and up the outside of the calf to the back/outside of the knee, where you can feel a tendon (Fig. 3.2 page 49 and the exercise on page 75). This is part of the outermost hamstring muscle,

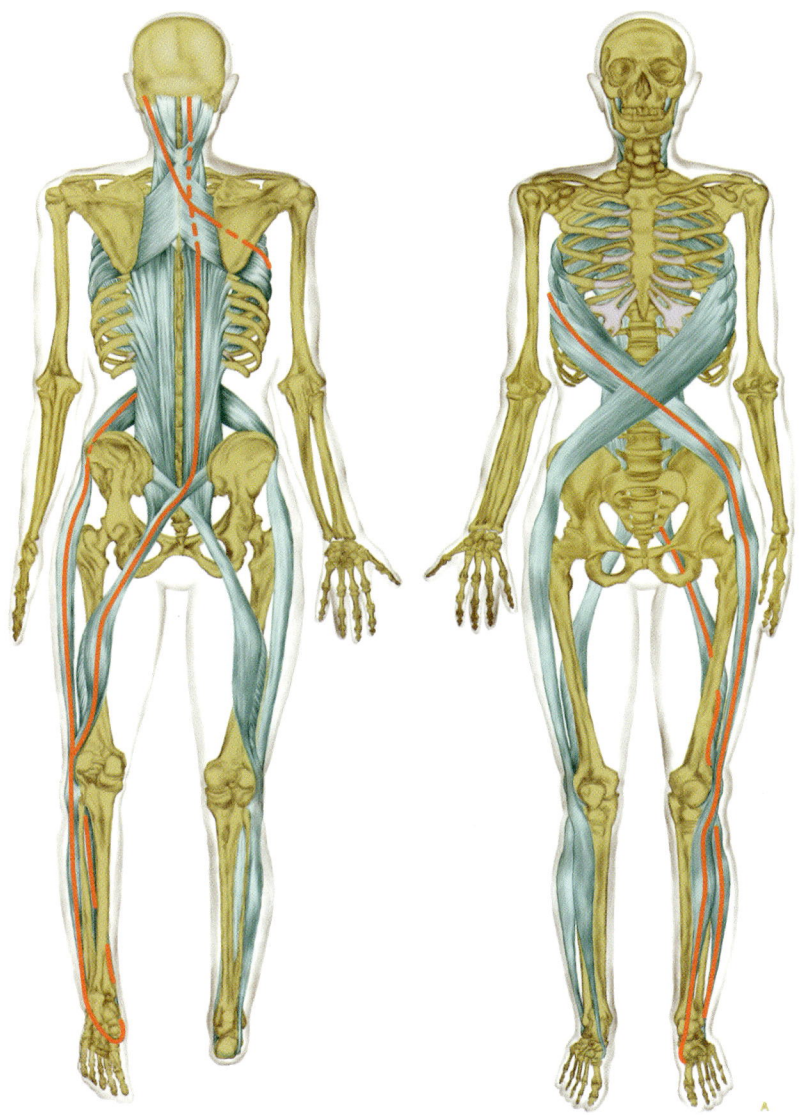

Fig. 8.1 The Spiral Lines from front and back. The red line highlights one arm of the 'X' on the back between the shoulder-blades, the 'X' across the abdomen, and the 'X' across the sacrum.

which takes a diagonal path to the seat bone. Just above this, at the sacrum, the line *crosses to the other side* (via the sacrotuberous and dorsal sacroiliac ligaments). These are embedded in the thick fascia that covers the sacrum and anchors the long back muscles going up this side of the spine. We have met these before as part of the Superficial Back Line, and they continue up into the neck, inserting into the skull on the *opposite* side from the starting point of the line, but slightly closer to the mid-line. Compared to the front back and sides, this is a complicated route; but we have made it all the way from one end to the other, and we are still alive!

Taken together, the left and right SPLs form a helical double lattice, crossing the mid-line to form three 'Xs' (Fig. 8.1). The top 'X' crosses on the upper back above and between the shoulder-blades. The second 'X' has its cross on the front near the belly button, and the third 'X' lies over the sacrum.

The SPL is central to proper shoulder-carriage. It passes from the spine to the shoulder-blade through the same muscles as the Deep Back Arm Line, but then continues *under* the shoulder-blade rather than over it, creating a sling for the scapula to sit in. The Deep Back Arm Line and the SPL are melded together, so that imbalance or hypertone can affect your ability to connect your arms into your back, and to push your hands forward.

Both the SPL and the Front Functional Lines of Chapter 6 make an 'X' on the belly, but not at the same place. Most riders enjoy finding the 'X' of the Front Functional Lines, and relish in the gains in their riding. Adjusting the wide muscular 'X' of the SPLs near the umbilicus involves deeper muscles and more sweat. While, at the beginning, this can create a much bigger feeling of 'weird', this 'X' talks to the heart of our asymmetry more effectively than any of the lines we have met so far.

Often when teaching, I have watched a rider adjust her SPLs and make stunningly effective changes to her asymmetry. Within moments, her new-found ability to steer changed her horse as well. with an understanding of the Anatomy Trains lines – and this 'X' in the SPLs in particular – you can quickly change what would otherwise have taken years of 'dripping water on the rock'.

Squeezing the Toothpaste Tube

If the SPLs are both slack in the upper body, they accommodate a forward head position and the rounded upper back of a 'dowager's hump', as well as a protruding belly. The rider slouches back and drops behind vertical; her mid-back looks hollow

Fig. 8.2(a) To mitigate the 'leakage' of the Spiral Lines, think of a connection through you from your upper chest to your spine, either to the level of your bottom rib, or to the top of your pelvis. Use this to help you keep a box-shaped torso as in (b).

(a)

(b)

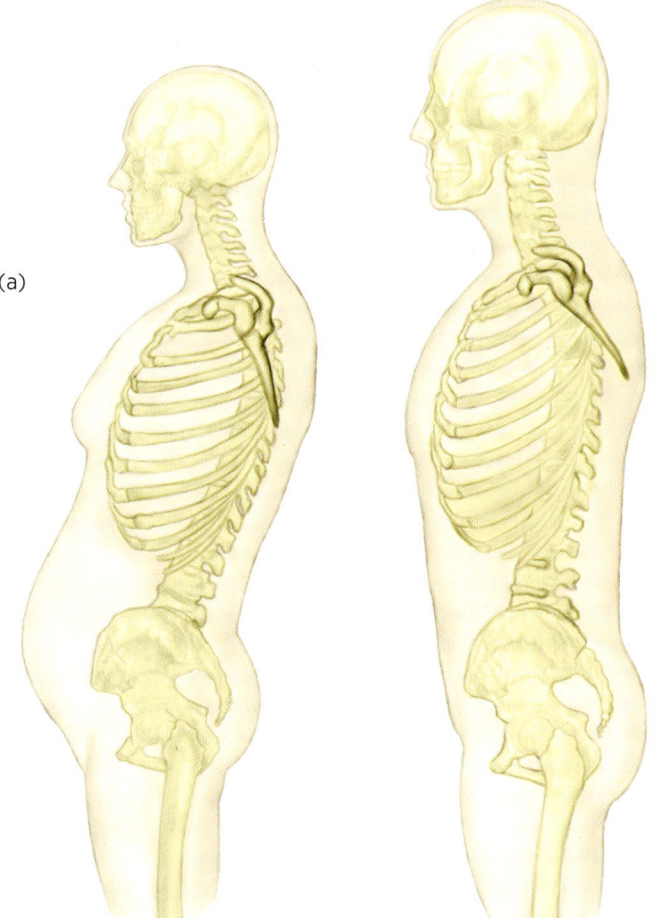

Fig. 8.3 The skeleton mirroring the postures of 8.2(a) and (b) shows how the shoulder-blades, which appear flat in (a), are not flat once the torso is repositioned. This highlights the need for the correction of 8.2(b). Notice that in both Figs. 8.2(a) and 8.3(a), the collarbone is virtually above the small of the back. But, in effective riding, the front of the torso-box is one vertical line, parallel to the back of the torso-box, which is a second vertical line.

and her belly protrudes forward. From the side, the torso thus makes an 'S' – a more complex pattern than either a hollow back (short Superficial Back Line and longer Superficial Front Line) or round back (short Superficial Front Line and longer Superficial Back Line, see Fig. 8.2(a)). This SPL 'leakage' adds the tenth option (j) in Fig. 3.5 page 57. We have already seen how it also affects the shoulder-blades, and see it here again in Fig. 8.3.

How to avoid this leakage:

1. Whilst sitting in a firm chair, imagine this 'leakage' in your body: round your upper back whilst sticking out your stomach. Some people can even manage to put their collarbone, which should be vertically above their pubic bone, so far back that it is almost directly over the small of their back! Put one hand on your upper chest and the other in the small of your back: do you come close to this?
2. Keep your hands in place as you make your front and back parallel and vertical, lifting the hand in front and using the hand on your back to ease it back even more. Imagine a diagonal connection inside you that goes from the notch at the top of your breastbone to your lower spine, joining your hands. Use your breath to keep your collarbone moving forward and your lower ribs back, so that the front and back of your torso-box stays aligned vertically (see Fig. 8.2(b)).
3. When you are riding, add this diagonal to the idea of 'resist my push' (see Fig. 3.7 page 61).

Side Bends and Lateral Shifts

As with all of the lines, the SPLs can pull from either end, or from nearly any place along their length. When they are imbalanced they 'participate in creating, compensating for, and maintaining twists, rotations, and lateral shifts in the body'.[1] Sometimes these imbalances are rooted in uneven tone in the SPLs themselves, but often the SPLs are compensating for a strong bend or rotation in the spine. Most of us have a slight bend or rotation in the spine of 4 degrees or less – though not ideal, it is very normal. Those with 20 degrees or more can be told they have 'scoliosis' – but it is a matter of severity, not pathology.

The new term in the description above is 'lateral shift' – although we encountered it before when we talked of one hip having a more 'girlish' cast than the other. When the ribcage moves laterally relative to the pelvis, it is necessarily accompanied by a sideways 'S' curve in your spine.

1. Think of 'break dance' and make a lateral shift of your ribcage to each side. Feel which is more familiar. Look in a mirror and see which side goes further (see Fig. 8.4).

Fig. 8.4 A lateral 'S' curve shift of the ribcage, shown here in leg-yield. In lateral movements this distortion is much more likely than the 'C' curve often seen on a circle.

2. If your displaced ribcage makes you lean in when riding, you will, if anything, have creases on the *outside* of your torso about a hand's width beneath your armpit (at bra level). The shift may also make you lean out, with creases under your inside armpit.
3. Make a lateral shift to the more familiar direction, and then add a side bend that goes the opposite way, lifting the shoulder that sticks out to the side.
4. Next, add a rotation that advances that same shoulder. You have done this before in the exercise on page 100, and the pattern is very common – how easily does your body fit into it?
5. For an experience of being 'not you', mirror-image this, beginning with a lateral shift to the other side, then the side bend, then the turn! Can you imagine this 'correction' becoming more familiar, less strange?
6. Once you learn to fix your 'C' curve, you may be left with the 'S' of the lateral shift. This is the problem beneath the more obvious problem! Also, riding shoulder-in and half-pass tends to induce lateral shifts rather than 'C' curves in the rider. The challenge of these movements is that forces are sent *diagonally* through the horse's the rider's bodies. Fig. 8.4 shows the typical result.

It is effective to work with the SPLs in sections. If we allow plenty of time to get a felt-sense of each part in practice, we can influence them to great effect. Within the torso, we will think of each 'X' in turn, and will then consider how the lines work within the legs.

The Upper 'X' – Easing Your Head-carriage

How can the SPLs influence head position? The vast majority of riders carry their head too forward and, when riding on one rein, they tilt it, looking down to the ground on the inside of the circle. Their outside shoulder is lifted and/or drawn forward, and their torso follows their head as they rotate and collapse to the inside. Their unequal SPLs, Functional Lines, and Lateral Lines all work together to create their 'bad' rein. They rotate to the inside as they pull on the inside rein, and… you already know the rest of the story, along with the price they are paying for their asymmetry.

Realise that what we do *on* our horses we are also doing *off* them. We sit at our computers pulling our heads forward, and developing the slump that is compounded by ageing. In many of our daily activities we are lengthening the upper 'X' of the SPLs (and also the underlying Superficial Back Line). Unwittingly, we gravitate slowly towards that walking frame!

In addition, we unknowingly build our rotational preference into how we set up our working environment. Our asymmetry often determines where we position our computer screens and favourite gadgets, and even where we sit around the dinner table in relation to our loved ones. In these contexts, such everyday compensations do not seem to damage us; but, over time, our distortions will lead to thickening within our sinews that will one day give us pain. On the horse, however, even small compensatory patterns work against us – whether we know it or not.

To these baselines developed off-horse, add the tendency to look down at the horse's neck whilst riding. Dressage riders in particular want to see how their horse's neck is positioned, and they often feel insecure if they cannot do this – they do not trust their felt-sense to give them accurate feedback. They tend to drop their head as well as their eyes and, in my experience, telling people to 'look up' rarely works for very long. If forced to look up some riders may literally feel unsafe, and may indeed be less able to process information. (See *For the Good of the Rider* for a discussion of this).[2]

At the right time looking up yields huge benefits, but until that time instructors can start to sound like a broken record!

1. As you sit in your chair, jut your head forward and look down. Realise how you have extended the upper legs of the SPL 'X' (the splenii muscles). Exaggerate this even more and widen your shoulder-blades to form a 'widow's hump'.
2. By far the most effective way I have found for getting riders to lift their chin, and put 'the back of their neck against their collar', is to think of adjusting the upper legs of this 'X' by letting their head float up and back.
3. Feel how this draws the back of the head back towards the centre of the 'X', reducing the hump in your upper back! Think of shortening the lower arms too, so that your upper back flattens.
4. Do you look down at your horse's neck when riding? Ideally you would be able to see the brim of your riding hat in the upper part of your visual field, his neck in the lower part, and the riding arena in the middle. This uses your unfocused peripheral vision – you are seeing everything whilst focusing on nothing, and letting the world come into your eyes. (You experienced this after the balloon breathing exercise on page 63.)
5. Can you divorce your eyes from the horse's neck enough to experiment with using a wider vision as you ride? If you are addicted to looking down, 'divorce' is probably the right word!

If I ask a student with head-forward posture to lean forward more (say in rising trot) she will often *thrust her head forward even more* in response. This rider has not changed the angle of her upper body, but is convinced that she has done right thing! If you were riding in the above example, I would then make sure that you understood – and could demonstrate – that leaning forward happens from your hip joints, not your neck!

1. Continue our experiments with the upper 'X' by tilting your head to one side. This almost always involves looking down: which side would you choose? Test both to find which is more familiar. Realise what this does to both sides of the 'X', including its lower legs.
2. Trace each of the four legs of the 'X' in your mind. You will probably discover that your normal pattern has two shorter legs on one side of your body, and two longer ones on the other.

3. The 'X' has created a 'C' curve that puts your head closer to your shoulder on your shorter side, and further from your shoulder on your longer side.
4. It is possible, but less likely, that your neck and ribs curve in opposite directions, making a sideways 'S'. Some riders develop this more complex pattern in a misguided attempt to fix the more simple 'C' curve. Make a 'C', and then imagine that you have been told to lower the ear that is higher!
5. Take plenty of time to keep analysing your own pattern, so that you know which legs of the 'X' are longer and which are shorter. Shake yourself out and begin again if you need to. Let your head float up and out of any curve you have until you approach 'equipoise'.
6. In the 'C' curve, notice how your ribs compress together where the shorter leg of the 'X' passes around your side, on its way to becoming part of the lower 'X' on your front. If you do not feel this, compress them deliberately.
7. Once you can perceive this 'X' in its entirety, notice where its centre would be relative to your spine. Can you bring both the centre of the' X', and your spine, onto your mid-line? This is a far better way to undo the upper 'C' curve than simply tilting your neck the other way!
8. How does this change your head, and what differences can you sense on each side of your spine, and in each side of your ribs? Can you equalise the four legs of the 'X'?
9. Repeat this a number of times so that you get clearer on the difference between getting it, and losing it. As you adjust your head and the centre of the 'X', can you 'push stuffing' into the ribs that were squished, whilst squishing the ones that were held up and apart?
10. Repeat this exploration whilst riding.

Working this way on the symmetry of the upper 'X' makes significant differences to riders. The changes can travel down the length of the SPLs, changing the rider's overall symmetry and ability to steer.

The 'X' in the Abs

The SPLs form an 'X' like the cross of St Andrew, from the side of the left ribs to the right hip and vice versa. This challenges us to think about the various pulls that act on the four legs of the 'X': these would pull your points of hip *and* the sides of your lower ribs towards or away from your belly button. Imbalanced pulls could even cause your belly button, as the centre of the 'X', to migrate away from your mid-line!

The following exercises are complicated. Be patient, go slowly and trust your felt-sense. This first exercise introduces you to the 'X'.

1. Think of 'break dance' and make a lateral shift of your ribcage. This will most likely make two legs of the abdominal 'X' that are diagonally continuous (the external oblique on the upper leg coupled with the internal oblique on the lower one) *longer and stronger* than the opposite two, which will feel short and fuzzy

by comparison. Can you make the stronger lines really clear? Put your hands on them if you need to. You may find yourself bearing down in the process, and this entire experiment could feel like a lot of effort.

2. Make a lateral shift the other way. Can you reverse this pattern, making the opposite diagonal longer and stronger? Take time with this. Which diagonal is hardest to get a sense of?

3. Make a lateral 'C' curve. Which legs of the 'X' are longer? I would expect it to be the two on your longer side, but they may well feel less distinct than they felt in the lateral shifts. Can you make them more distinct to your felt-sense?

4. Make a 'C' curve the other way. Can you make the longer, stronger legs really distinct to your felt-sense?

5. Can you make the bottom two legs of the 'X' strong and symmetrical?

6. Can you make the top two feel strong and symmetrical? In order to do this you may have to really advance your ribs (rather than lifting them), and pull your stomach in *more than you have done until now* as you make it into a wall.

7. It may help to think of this shape > < as you attempt to get all four legs to feel more equal and functional. Can you get anywhere close to this? What would be your default pattern – do all of the legs become fuzzy? How much can you change this?

8. Have any of changes you made in this exercise affected the evenness of your 'stuffing' in each side of your torso-box, and the tendency to make creases in one side of your torso?

The Heavier Seat Bone

In this exercise, assume your body makes a 'C' curve one way or the other, as most of us do. This exercise shows why the inside seat bone usually becomes heavier on the rider's bad rein, with the outside seat bone floating off sideways – perhaps to the point where it comes off the saddle. Or you may belong to the much smaller percentage of riders who compensate by putting more weight on their *outside* seat bone.

1. As you sit in your chair, let your body fall into its favourite 'C' curve, and notice the 'X' across your abdomen. Trace its legs in your mind. Are they all quite fuzzy?

2. Now think about the bottom two legs of the 'X', and exaggerate the difference between them by elongating the longer line. Does it come naturally to do this by *drawing the front of your hip down away from your belly button, or your belly button away from your hip?* Try it both ways, tracing the line in the appropriate direction in your mind, and maybe with your hand if you have a hard time sensing it. How does your torso distort in each case?

3. Now shorten the shorter line even more. Do you instinctively draw *your point of hip towards your belly button, or your belly button towards your hip*? Try it both ways.

4. When you draw your point of hip towards your belly button you will feel your

seat bone on that side lift. Meanwhile, your backside on the opposite side drops down and bulges outwards, becoming more 'girlish'. You have a larger surface area down on the chair, with your seat bone feeling heavier.

5. Now draw your belly button toward that same hip. Feel how this makes you *lean* towards the same side, as this seat bone becomes heavier and the other one lightens. Recognise that the concept of 'shortening the muscle' is not enough to determine your fate – this depends on *how this increased tension pulls on your skeleton*: one way makes the seat bone on that side heavier, whilst the other way makes it lighter!

6. For the vast majority of people, trouble arises when the shorter lower leg of the 'X' (the internal oblique) is on the inside. This means that the butt cheek on the side that is longer is on the *outside* when riding on their 'bad' rein.

7. When riding, most people shorten the short line by *drawing their belly button towards their point of hip*. Even if the other way comes naturally to you when sitting in a chair, you may well find that you do the opposite when you are riding on a circle and affected by the centrifugal force.

8. Drawing your belly button towards your hip makes you lean to that side. You have weighted what would be the inside seat bone on your 'bad' rein. The outside seat bone will tend to slide off the side of the saddle as your shoulders lean to the inside. This is the pattern of most riders' asymmetry.

9. Repeat what you did in 2 above. Make your torso into your favourite 'C' curve by lengthening the longer bottom leg of the 'X'.

10. Now for the toughest but most effective fix: think of shortening it. Do you need to *draw your hip diagonally towards your belly button, or your belly button towards your hip?* It will probably be the former when you are riding, if not in your chair.

We are now more than halfway there, and if you have followed each step in this, well done (especially if you do not have the most common pattern)! Realise how much body awareness you are building, and how valuable this will become.

I am sure you will not be surprised to realise that our next step is to think about the upper two legs of the 'X' (the external oblique muscles) finding a way to even them out too.

11. If you move your awareness from the longer bottom leg of the 'X' to its diagonal continuation, this upper part probably needs to *lengthen* as it extends towards your opposite ribs. (Again, trace the line with your hand from your belly button towards your ribs.) You are searching for an adjustment that will level out the underneath of your torso-box and shorten your longer side, which might even begin to feel as if it has creases in it. Concurrently, you will fill out and lengthen your shorter, soggier side.

12. This might feel like the opposite lateral shift to 'home'; it is unnatural, and a huge amount of work. But it really will help to mitigate your inherent asymmetry, as well as the way it is triggered by the centrifugal force on a circle. When one seat bone becomes heavier when you ride, this is your best shot at a fix.

13. Now think of elongating the lower leg of the 'X' that is shorter, from your belly button towards your hip bone. Additionally, its extension towards the ribs on your longer side needs to shorten, pulling your ribs towards your belly button. This too will feel weird.
14. Can you make each of the sections of your rearranged 'X' firm and clear to your felt-sense? Make sure you are breathing!
15. As you even out all four legs of this 'X', repeat the process with the upper 'X' on your back. Can you stack the centre of the upper 'X' above the centre of the lower one? Feel how much stability you have produced in your core!
16. You now have a way to keep your underneath much more level on the saddle, and the two sides of your torso-box much more even and vertical. Once you can do this effectively enough in motion, you can wave goodbye to that 'C' curve as it rides off into the sunset!

The 2.0 Problem

Riders who have worked a lot with an asymmetric 'C' curve can be left with a few creases in their ribs about a hand's width below the armpit (at bra level), just where the lower leg of the upper 'X' wraps around their side (with the serratus anterior muscle). This gives them a lateral shift in their ribcage – an 'S' rather than a 'C' curve in their torso.

I often see this in leg-yield and shoulder-in, when the rider attempts to move the horse over. A desperate frenzy of kicking and shoving maximises the problem, but even relatively skilled riders are not immune. The rider's shoulders are easily left behind the ideal vertical balance point, as her torso deforms into that sideways 'S', and creases form beneath her armpit on the side she is moving away from (see Fig. 8.4 page 164).

Sometimes I see similar creases on the *outside* on a circle when the rider wants the horse to move away from that side. Usually she also has the outside rein pressed against the horse's neck in an attempt to turn him.

When riding on a circle, this rider now *actively recruits* her asymmetry as a turning aid when the shorter side of her body is on the *outside*, and she is riding in the *opposite* direction from the original 'bad' rein! Her body has found a sneaky new use for her (much-reduced) asymmetry. She is no longer at the whim of the centrifugal force, which caused problems when her shorter side was on the inside of the circle.

Many riders never evolve into this '2.0' version of this problem, but if you find you have, know that you have made considerable progress! You have peeled your way into a new layer of the onion and, fortunately for you, the fix you need in this new phase in your learning will be much, much easier than the fix for the original problem!

1. Make a lateral shift of your ribcage (see Fig. 8.4 page 164). Feel how your ribcage fills out at about a hand's width beneath the armpit on one side, whilst scrunching on the other. Make a lateral shift in the opposite direction. Realise

Fig. 8.5 Strengthen the outsides of your torso-box by pushing your knuckles together.

Fig. 8.6 Fill into the firmer edges made in Fig. 8.5 by attempting to pull your hands apart when the fingers of each hand are curled around each other.

that what feels like the opposite shift may, in reality, be a vertical spine! Check this out in a mirror or via a friend with a camera.

2. Imagine a bolt through you at a hand's width beneath the armpits, joining your two sides together and strengthening your torso-box.
3. Tightening a nut on this imaginary bolt gives you a way to pull your sides in towards each other, giving you stronger 'walls' in a box that is slightly narrower. Then push out against those walls without letting them give way. Figs. 8.5 and 8.6 show how resistances with your hands can help you do this.
4. I hope this idea has a familiar ring to it: you are now bearing down in your sides, and engaging the muscles of the Lateral Lines as well as the SPLs.
5. Can you turn your horse whilst keeping this bolt, and keeping your outside rein away from his neck? Does this help you arrange your 'double yellow lines' over the horse's 'double yellow lines'? (See page 119.)

The Spiral Lines in the Legs

It is now time to think about the SPLs as they wrap around each leg. Trace the line with your fingers (see Fig. 8.1 page 161), beginning from the front point of your hip bone. Find the muscle just down from its outside edge, and track it down the front/outside of your thigh toward the outside of your knee, where you can feel the strong tendon of the iliotibial tract. This connects *across* the front of your calf (via the tibialis anterior), arriving just in front of the bony knobble of your ankle, and thus to your instep.

From under your foot the line passes just behind the outer knobble of your ankle, and up the back/outside of your calf, past the back/outside of your knee, and across the back of your thigh to your seat bone. Repeat this until you are so familiar with the line that, whilst riding in walk, you could trace it in your mind, and perhaps trace parts of it with the knob end of your stick.

Once you can do this, you might well find that parts of the line – especially in your calf and/or outer thigh – feel really fuzzy. You might even feel as if parts of it are floating around outside of your body, weird though that sounds! Can you somehow find a way to draw these parts into your body, so that the entire line functions within the boundary of your skin? If you can effectively 'parcel' your leg within the line, you will gain more control of your lower leg, and a sense of your legs being *longer*. This adds to the improvement that comes from continuing the Functional Lines down to your feet.

When the SPLs are not working well in the legs, they may pull your knees and ankles out of true. You have probably had the experience of putting on someone else's shoes that are the same size as yours, but feel very strange. The different wear pattern is created by differences higher up in our various lines of pull. When the SPLs are more even around your legs, you are more likely to be able to spread weight evenly across the balls of the feet, with the heels more out and the toes more in.

1. Stand with your feet parallel and slightly apart, and your pelvis tucked under – known as the 'horse stance' in martial arts.
2. Slowly bend your knees, and look down – do your knees follow a line directly over your feet, or do they track instead to the inside or outside? If you cannot see your big toes as you flex your knees, they are falling to the inside – keep the kneecaps going right over the second toes.
3. In standing and riding, do you tend to drop the inner arch of one or both feet, with your toes turning out? If so, an observer will tend to see the sole of your boot when you ride. Or do you weight the outside of one or both feet, along with your little toe? Whichever tendency you have, your aim is to bring the arches and your weight distribution to neutral. Resisting the heels out (as in Fig. 5.17 page 117) can help to develop this, as can resisting inwards. Whoever gives you these resistances needs to observe carefully that you do not roll your ankles either way. Be mindful of this yourself.
4. How clear is your sense of the SPLs within your back outer calves? Reach down to the outside of each knee and strum across the tissue: the SPL includes the tough tendon right on the outside, which connects into your outer hamstring.

Repeat the exercise of page 75, and make that tendon stick out against your fingers. If you are struggling, point your toes out to make it clearer. Then bring your toes back to the front, keeping the tendon sticking out.

5. Can you clarify in your mind how the SPLs pass through your outer hamstring muscles to your seat bones?
6. How do each of these steps change your perception of, and control of, your legs and feet?

When you succeed in wrapping your legs within the SPLs you are likely, in addition, to suddenly become aware of your big toes! This is not so strange as it sounds. The 'stirrup' where the line passes under your instep is made of two tendons that both connect into the long bone that becomes your big toe. Thus tensioning the SPL within your leg activates your big toe.

The Sacroiliac 'X'

When the SPLs arrive at the sacrum they cross, for the third time (see Fig. 8.1 page 161). Thinking of this 'X' helps you become narrower in the back of your pelvis, enhancing the narrowness of 'both boards on' (the Intermediate Stability System see page 107). This in turn helps you to keep each seat bone on the inner edge of the horse's long back muscle (which is the outer edge of the gullet of the saddle), making it much easier for you to organise both sides of your horse. Pulling outwards on the gullet of the saddle, as in Fig. 8.7 page 173, can help you narrow in.

The SPLs then follow the same path as the Superficial Back Lines, running up the long back muscles (but now on the opposite side from the leg stirrup) until they arrive at the ridge on the back of the skull. It can help to think of the narrowness at the sacrum continuing all the way up the back – I have known riders become very excited about the changes they make to their back and their horse's back if they imagine a can-can corset with laces going from their tail bone to the nape of their neck! Tighten that corset to bring your boards together.

We have now traced the entire length of the SPLs. All the switching from right to left and the complex route of these lines makes them hard to follow, but absolutely essential in riding. If it was too much to take in, come back to it when you are ready for the next level of challenge. Sometimes improvement comes as a little bit here, then a little bit there, and then some more back here again – the Spiral Lines especially are like that.

Fig. 8.7 Pulling outwards on the gullet of the saddle can help you learn how to transmit force across the bottom 'X' of the Spiral Line. It also helps to narrow in the back of the board on that same side. You can also pull on a strut on the back of a hard chair.

The Spiral Lines in the Horse

The SPLs have a long way to travel (through much longer splenii muscles than in humans) from the horse's poll to just in front of his withers and their first crossing point. From here, each line passes under his shoulder-blade, and curves down around his ribcage towards the second 'X', which is back by his umbilicus. Each SPL then travels up to and around the point of his hip before going down the outside of the thigh to the hock (our heel). It wraps around the front of the hock before travelling up the back of the hind leg through the Achilles tendon and hamstring muscles. After their final crossing point on the croup, the SPLs follow the path of the Superficial Back Line, returning to the horse's head on the opposite side from their starting point (see Fig. 8.8).

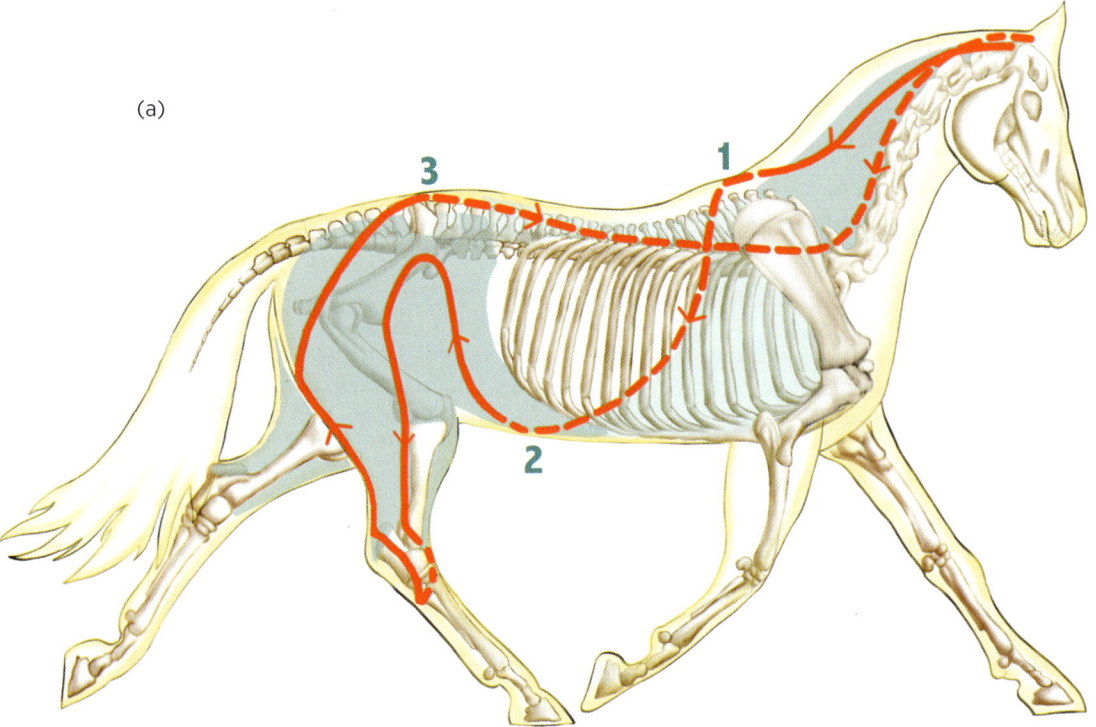

Fig. 8.8(a), (b) The Spiral Lines in the horse, from the side (a) and above (b). The red line is dotted on the far side and underneath of the horse. The arrows help you trace the line, but remember that it can pull in either direction from anywhere along its length. The three 'Xs' where the lines cross on the horse's mid-line are numbered: (1), just in front of the withers, (2) far back on the stomach at the navel, and (3) on the croup. The complexity of the lines allows for tilted muzzles and necks, leaning shoulders, and rotated ribcages … some of the trickiest problems for the rider to fix!

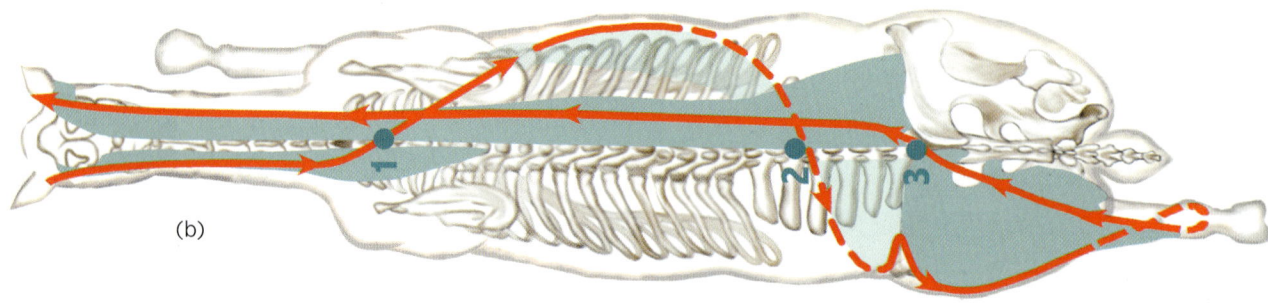

Like you, the horse has many ways of distorting the 'X' on his belly, and unequal SPLs can effectively wring his body out like a dishcloth. We have already seen how the resulting torque between one shoulder and the opposite hip often underlies our asymmetry. If your horse's rugs continually slip, you now have an idea of the prime suspect – but remember that none of the lines ever work in isolation.

It is the SPLs that allow bend to happen in its ideal coordinated form: this requires the belly 'X' to have two longer arms on one side, and two shorter arms on the other. The ribcage rotates, which shortens its inside and fills out its outside, creating that reach into the outside rein. The withers 'X' also plays its part, creating a slight tilt in his central axis. As we shall see, 'bend' requires the whole of his body to lean slightly in – and the Lateral Lines alone cannot make this happen.

Surprisingly, the horse's long back muscles stay remarkably level and even, with the ribcage rotating underneath them. Indeed, as each SPL passes along the top of the body (following the line of the Superficial Back Line), it limits the extent to which the lower part of the line can affect the shoulders and pelvis. The SPL also flexes the horse's torso to raise his back – and the undulations of canter, especially, utilise these healthy movements of rotation and flexion.

You already know that the horse does not have collarbones, so his shoulder-blades float in the soft-tissue lines without an attachment to the rest of his skeleton, and move by gliding over the underlying tissue (as do our own, when they are free). The horse's ribcage is literally *suspended* between his shoulder-blades and upper arm via the 'myofascial sling' (see Fig. 8.9(a)).

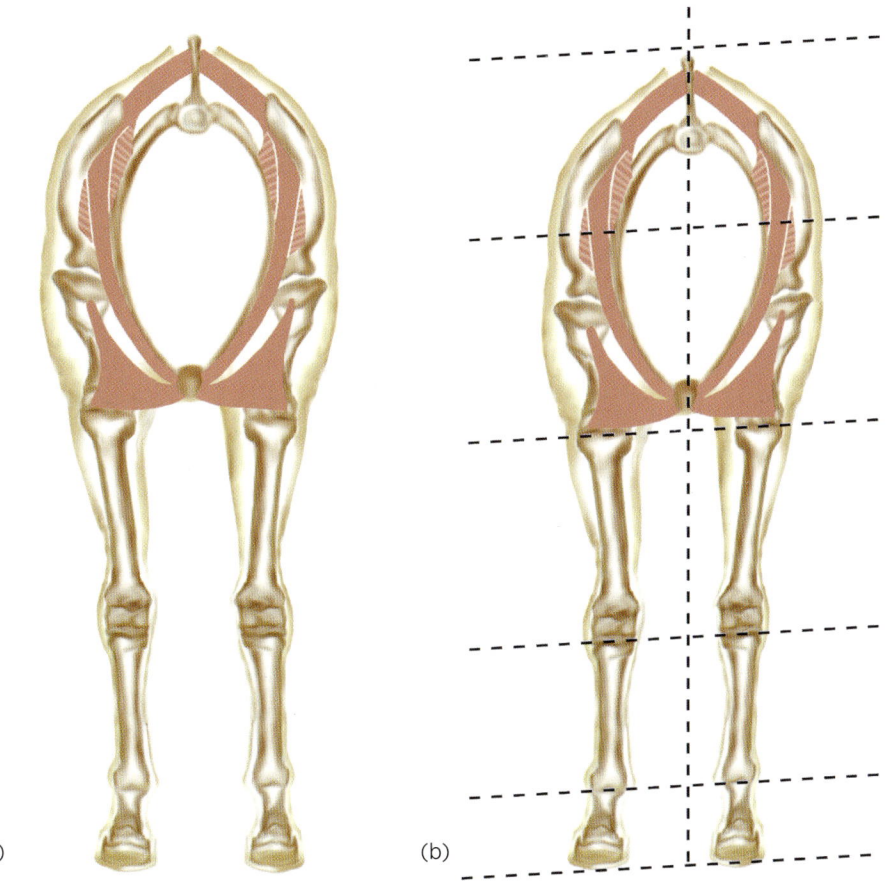

Fig. 8.9(a), (b) The Spiral Lines form part of the muscle sling that suspends the horse's ribcage between his shoulder-blades, passing through the rhomboid muscle by the withers, and the serratus muscle, which is shown stripped. In (a) the horse is straight. To understand the horse's contortion in (b), playact it in your own body, placing your arms as if they were his ribcage: he has leant his withers to the left, and then used the muscle sling to tilt his ribcage so that it stays vertical. The forelegs are then neither vertical nor level … and the knock-on effects throughout the various fascial lines are huge. In motion, we perceive these tilts in the neck, muzzle and ears: the right ear would be carried lower.

This arrangement also gives the horse's ribcage freedom to swing between the shoulder-blades. In contrast, the movement of our human ribcage is restricted by our upright spine and our collarbones, which are like struts that hold it in place. If we crawl on all fours, we cannot imitate the freedom of the horse's ribcage to swing within the SPLs, or to find the axis that is needed in 'bend'.

Dr Deb Bennet, who proposed the above comparison,[3] is one of the most profound thinkers on the subject of straightness, and she defines a (standing) straight horse as having his sternum centred between his elbow and shoulder joints (see Fig. 8.9(a)). However, in one of the ultimate contortions, the standing horse could (for instance) lean his torso to one side, and then use his SPLs to rotate his ribcage and keep it vertical! This puts his sternum closer to one foreleg, so that a plumb line dropped from his sternum lands closer to one hoof. His forelegs then slant, their joints unlevel (see Fig. 8.9(b)). God help both rider and farrier!

These kinds of contortions become clearest to us riders through the tilt they create in the horse's neck and muzzle; so as the withers rotate to one side, the ear on that same side is lowered. The rider simply lifting the hand corresponding to the lower ear will not make a viable long-term fix; instead both SPLs need levelling out – a big challenge for the rider, but potentially possible!

Whether the horse's muzzle tilts or not, sensitive riders often realise that the back end of the horse has tilted, putting one 'back corner' of their torso-box on a surface that slopes diagonally down and back. Either the rider's underneath slopes away with the horse, or she holds her body level in a way that levels him through his body. This is skill in action!

Bear in mind, however, that hind leg lameness could also be creating this feeling, so there is a grey area in which the skilled rider might be able to 'ride a lame horse sound'. Alternatively, the rider's own asymmetry issues might have contributed to that lameness; we are left hoping that no long-term damage has been done, and that by fixing her own asymmetry, she can fix the horse.

And so We Finally Arrive …

Your aim is to learn how to ride your horse so that he develops two equal, level long back muscles. As you learn through Chapter 5 how to stack your 'double yellow lines' over them, you become more able to turn him 'like a bus', so that he follows your imaginary line with his withers. The 'A' frame of your thighs cuts down its 'wriggle room'; his and your Lateral Lines are become more equal in length, and your rein contact becomes more even. The 'train' of his body can then be pushed along from the engine in the hindquarters, without jack-knifing or 'derailing'. Then it is time for the next step… and thus we have finally arrived at the issue of bend!

We are working against the horse's instincts here, since a horse turning at liberty will lean in and fall in more than we riders appreciate as he jack-knifes to the inside. But realise that, left to themselves, horses rarely jack-knife to the outside and fall out – this problem is largely created by poor riding. Horses are subject to the same laws of physics as aeroplanes and bicycle riders, and recent veterinary research has

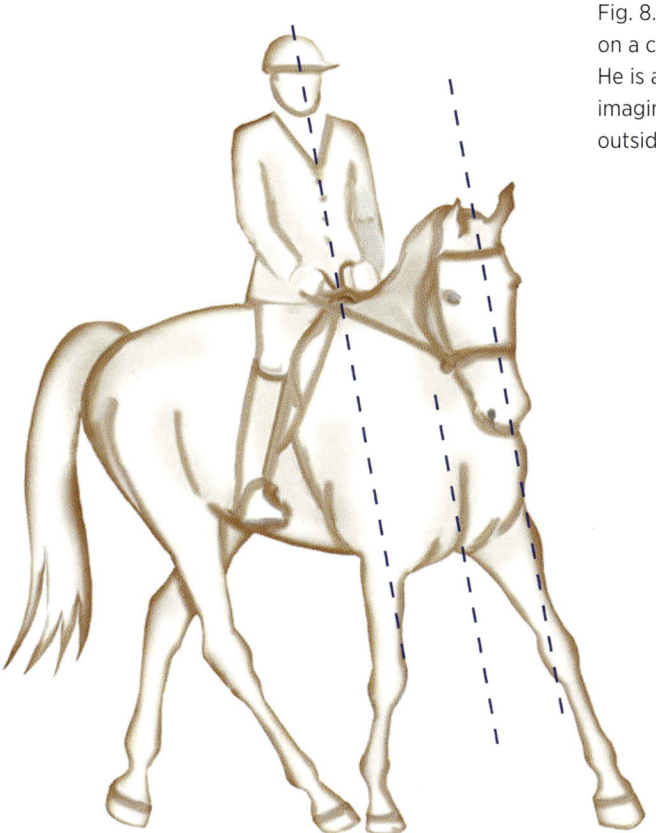

Fig. 8.10 Shows a horse and rider working well on a circle, tilting their central axes appropriately. He is able to 'bend' whilst following the rider's imaginary line, rather than jack-knifing to the outside or leaning in too much.

shown that well-trained horses lean in *less* than is predicted by the appropriate mathematical formula, whilst less experienced horses lean in *more*.[4]

Ideally, horse and rider tilt their central axes slightly (and we hope equally) on a balanced turn (see Fig. 8.10). Realise that the trained horse has *shortened* the inside of his body within that lean, whilst the untrained horse has *lengthened* it. Here is the key: in a balanced turn the shoulder-blades and ribcage *tilt as one unit,* so the horse does not get led astray by one errant foreleg (and its shoulder-blade, and that Lateral Line etc.…) heading off in its own chosen direction to create a jack-knife.

With the correct tilt in place, the SPLs and Lateral Lines allow the horse to bend around a turn, through tiny displacements in each of the intervertebral joints. The spinous processes of his vertebrae are slightly tilted, and if they and/or the muscle sling refuse to tilt that little bit, bend is not available!

Within this process, the ribcage is also drawn up between the shoulder-blades, lifting the withers and potentially facilitating collection – something that cannot happen in a jack-knife. The question for the rider is, 'How to do it?'

My answer is this: work on turning 'like a bus', and ride many small circles in walk, really paying attention to tiny deviations from your imaginary line, and learning how to fix them. When your horse can balance well (you can even think of him like a gymnast on a balance beam) you will feel that his withers and forelegs move in a very different way. The horse's axis has that tiny tilt as the SPLs work their magic; then, as you build your skill, you can gradually become more creative in the movements you ride.

Spiralling Around, or Not?

The directive we most often hear regarding steering is 'inside leg to outside rein'. As we have seen, being able to keep a contact on the outside rein requires precise positioning of your seat bones, thighs and torso, with your mid-line over the horse's mane and your skewers pointing along the tangent of your circle (see Fig. 5.7 page 103). Without this, you will inevitably give away your outside hand – despite all your best intentions.

The other common directive is to position your pelvis parallel to your horse's pelvis and your shoulders parallel to his shoulders. This is sometimes called 'the spiral seat'. These two directives represent some of the dressage world's most significant dislocations between expertise and explanation, and they get average riders into real trouble. The truth is that whatever theory you espouse, you will almost certainly obey it (if you ever actually do) on one rein and not the other. In real life it is your lines of pull that determine what happens when you steer – and so often they create a disconnection between your mental intent and its physical execution.

Both riders and ice-skaters cut their teeth on circles, learning how to work with the centrifugal force; so the ice-skater is analogous to horse and rider. The skater begins a circle by pushing off with the outside leg and then gliding on inside one, often whilst making a dancer's arabesque. Where does the skater's torso face during this process? It is to the outside, with her skewers pointing *even more out* than the

Fig. 8.11(a) Ice-skating position and (b) fencing lunge position seen from the inside of the circle.

(a)

(b)

tangent of the circle (see Fig. 8.11(a)). If the skater starts to twist to the inside, she will lose control of the turn, spiral outwards, and possibly even fall. This is the equivalent of the horse falling out on the circle.

This means that 'ice-skating position' (or 'fencing lunge position', Fig. 8.11(b)) is also an ideal for riders. To mirror it, your skewers may also need to actually point – or at least *feel as if* they point – further out of the circle than the tangent. This position enables you to produce a force from the outside back of your torso-box to the inside front of it.

But what if the horse falls in on a turn? His body is producing a force that sends his shoulders inwards, so the rider has to prevent him from leaning and falling to the inside. A different tactic is needed, requiring a force from the *inside back of the rider's torso-box to the outside front of it*. The rider is encouraging the horse to fill out the whole outside of his torso. It may also be necessary to anchor his hindquarters, since they could drift outwards as his shoulders fall in.

This realisation gives an overview that clears confusions, enabling us (in the next section) to flesh out the many partial explanations of steering. Realise too that the more the horse's body is *not* functioning as a balanced tensegrity structure, the less 'stuffed' he will be overall, so he will fall out at the weak points. The less well his lines of pull transmit force, the more 'drunk' he will feel! In contrast, the more tensegrity he has, the more connected and filled out he will naturally be – and this makes it very much easier for the rider to add to and adjust that filling, knowing that his body has more defined boundaries.

'Rebars'

Steel reinforcing bars (rebars) are used in concrete pillars – they are often visible on construction sites. We are going to imagine one of these lying just within each vertical corner of your torso-box, reinforcing it (see Fig. 8.12). Thus your 'rebars' are significantly further from your mid-line than your boards, and whilst they are not made of such strong strapping, they are immensely powerful. This exercise is best done in a carver dining chair, or one of those plastic garden chairs with arms.

1. Sit towards the front of your chair, in neutral spine. Reach your left hand across your back, and curl your fingers around an upright strut on the back of the chair. Pull on the strut, and think about your front left rebar – i.e. the one in the position that is diagonally opposite your hand. Can you make it firm up?
2. You need to be able to fill out your torso into this defined edge. Find a way to also *push* on the chair, and feel if you can fill yourself out into the front left corner of your box.
3. If you were doing this exercise on a saddle, you would take your arm across your back and wrap your fingers around the back opposite edge of the cantle (see Fig. 8.12(a)).
4. Now reach your left hand across your front and pull on the arm of the chair. As you do this, think of finding and firming up your *back* left rebar. If you do not have access to a chair with arms, use your front/outside thigh as a resistance for

the little finger end of your fist. When you are riding, you may be able to pull on the saddle cloth; however a bucking strap is a better option (and is worth having just for this exercise!) (see Fig. 8.12(b)).

5. Can you find a way to push against the arm of the chair (or your thigh, or the bucking strap) so that you fill out into that corner of your torso?

6. Sit for a while, and feel the difference between the sides of your box. The left side should feel defined, and make a clear edge. The right side will feel fuzzy and permeable. This exercise gives you a wonderfully clear sense of the difference between 'leaks', and boundaries. Firmer boundaries increase strength and skill.

7. Repeat the above steps with your right hand, both pulling and pushing first on the back of the chair and then on the arm of the chair. Remember that, whichever position your hand is in, you are searching for the rebar in the *diagonally opposite corner* of your box.

8. How has this changed the boundary on the right side of your body? Do you now have four clear rebars? Which is easiest to find, and which is most nebulous? Realise that this is determined by the fascial pulls which create your 'C' curve, rotation, and/or lateral shift.

Fig. 8.12(a) Bring your arm across your back and pull on the opposite side of the cantle. You aim is to firm up the front rebar on the side that your arm originates from – i.e. diagonally opposite to this hand position Then push on the cantle to fill into that rebar.

(b) Bring your hand across your front, and pull ideally on a bucking strap (or the saddle cloth). Your aim is to firm up the rebar in your back on the side that your arm originates from – i.e. diagonally opposite to this hand position. Then push on the bucking strap to fill into that rebar.

9. Practise this often whilst sitting in a chair, and then also whilst riding in walk.
10. Are you struggling to believe that elite riders have this much pressure in their torso, and this much firmness and precision in the corners and sides of their box? The best riders have at least some of the rebars some of the time – and would improve their skills by having more of them more of the time! However, they will probably tell you that they are 'just sitting'!

If you have the skills we have delineated to this point (which is no mean assumption!), then finding one rebar when you ride is better than having none. Having two is better than one. Three is better than two, and four is best of all. However, if you were suddenly to find them all, your horse might wonder what on earth had hit him! In reality, you will ease your way into this slowly, giving him time to adapt, and that is just as well.

Now we can apply this to circling. Sit on your chair, with all four rebars in place as best you can.

1. Think of the underneath of your torso-box. Can you make a diagonal connection within it, from the left back corner to the front right corner? This will make you more of a parallelogram with your right front corner advanced, and your left back corner more back.
2. Imagine your diaphragm as a cross-membrane within your box at the level of your lower ribs. Can you stretch it between your left back corner and your right front corner and make a connection between them?
3. Think of the tendon at the back of your left armpit, and the tendon at the front of your right armpit. Can you make a diagonal connection between them?
4. Sit like this for a while, and breathe. You are making a connection (and potentially a line of force) within your torso-box from the left back to the right front. This is a right fencing lunge, and the position which will bring your horse's shoulders to the right on the right rein.
5. However, should your horse want to *fall in on the left rein*, a force from the inside back of the box to the outside front of the box (and received by your outside arm) will help to keep him out on a turn. This will also create bend, filling him out to a defined boundary on his outside. I suspect that this 'rebar connection' is the true meaning of 'inside leg to outside rein'.
6. Now make the opposite diagonal connection at each level, from your right back corner to your left front corner. Your torso now makes the opposite parallelogram. Which set of diagonals comes more naturally?
7. Realise that this is fencing lunge left, and helpful when a horse wants to fall *out* on the left rein. The force from the outside back to the inside front mitigates this tendency, and also counteracts the effects of the centrifugal force.
8. This parallelogram will also serve you well when a horse on the *right rein wants to fall in*, and you have to mitigate this by making a force from the inside back corner of your torso-box to the outside front corner. This will help to keep him out and to create a bend to the inside.

9. Practise these positions in a chair, even if you are nowhere near being able to utilise them on your horse, and are currently more in need of the A, B, Cs of riding skills. They will help you mitigate your asymmetry, reducing any 'C' curves or lateral shifts. One day when you are riding along, you will suddenly realise that you actually have a rebar in place! Then start building on it!

There are a number of horses whose lines of pull send significant weight towards one shoulder. Whichever direction you ride in, you always need to orientate yourself to maintain *one* of these diagonals.

Leg-yield and Shoulder-in

We now have a wonderful way of thinking about a leg-yield with the horse's tail to the wall. This turns the horse's entire spinal axis by about 30 degrees. Shoulder-in would add a bend, whilst keeping the horse's hindquarters parallel to the track (see Figs. 8.13(a), (b)).

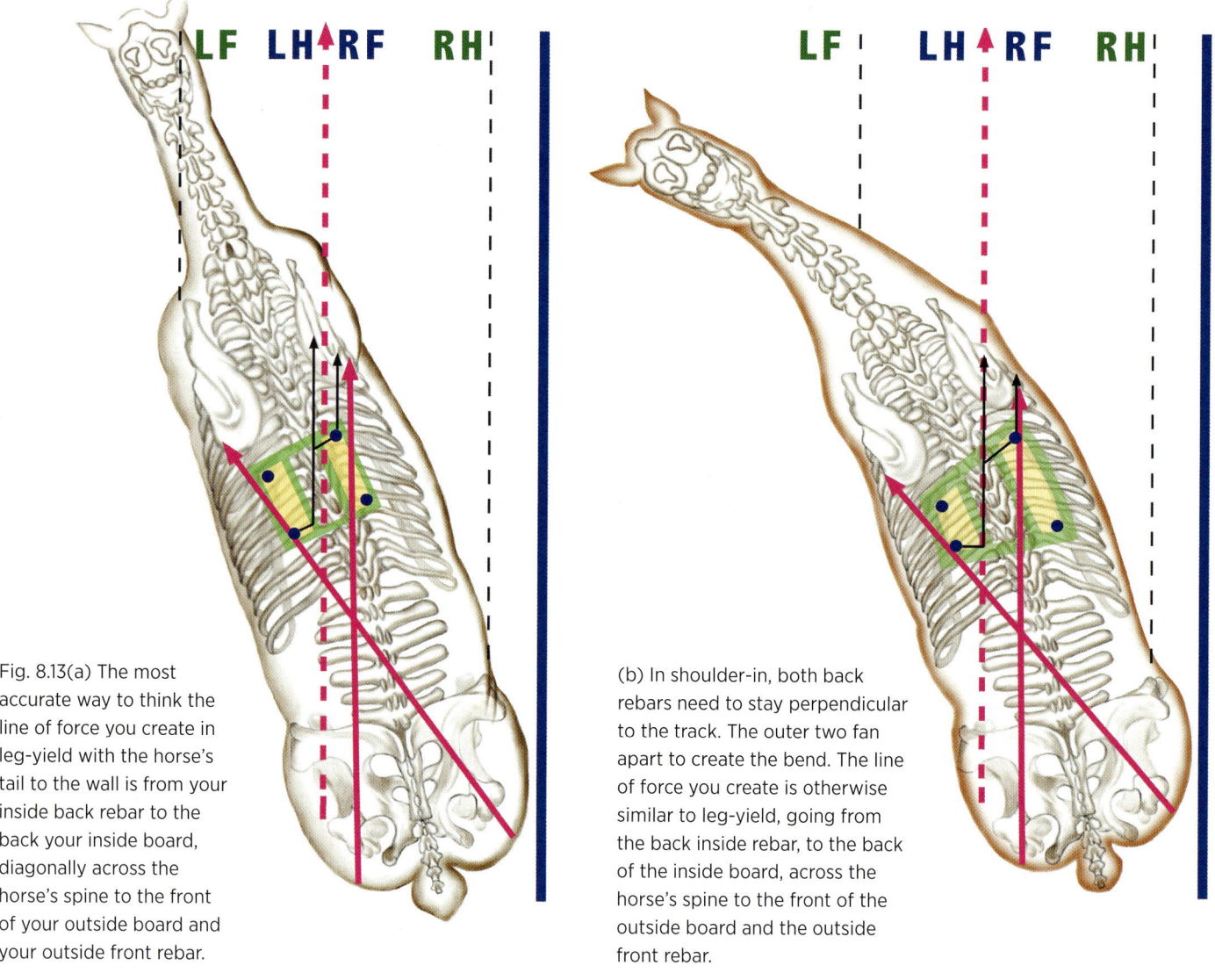

Fig. 8.13(a) The most accurate way to think the line of force you create in leg-yield with the horse's tail to the wall is from your inside back rebar to the back your inside board, diagonally across the horse's spine to the front of your outside board and your outside front rebar.

(b) In shoulder-in, both back rebars need to stay perpendicular to the track. The outer two fan apart to create the bend. The line of force you create is otherwise similar to leg-yield, going from the back inside rebar, to the back of the inside board, across the horse's spine to the front of the outside board and the outside front rebar.

If you could line up your inside back rebar with your outside front rebar, and keep the line that joins them parallel to the track, you would rotate both yourself and your horse about your central axes. This would position him into leg-yield! It is even more accurate to think of force transmission as shown in Fig. 8.13(a). Both boards and rebars need enough resilient strength to keep your torso-box intact – commonly, riders make contortions in their attempts to reposition the horse (like twisting the layers of a Rubik's Cube). Once the torso-box deforms, the horse will not rotate along with it, and then his 'resistance' contorts the rider even more.

To keep control of the horse's quarters when riding shoulder-in, you would need to add the outside back rebar to those two, and keep a line joining both back rebars *perpendicular to the track*. You then need to fill out the outside of your box, increasing the distance between the back and front rebar on that side, rather like opening a fan. This also happens on the outside of a circle – you and your 'bent' horse are no longer box-shaped! (see Fig. 8.13(b)). Shoulder-in can only work well if you can define and maintain the boundary on the outside of both of your bodies. You can then fill the horse out into this; but without it, he will merely hinge at the withers and jack-knife down the long side.

For many riders, learning to ride the lateral moments is like learning a new language. The movements invite riders to make a lateral shift in their ribcage as they think of 'pushing' the horse over (see Fig. 8.4 page 164). The seat bone, board, and front rebar in the direction of motion so easily go walkabout. It is unbelievable how counter-intuitive the correct fixes are, and how hard-won they are for most riders.[5]

'A waist is a terrible thing to mind' – for some riders at least. But if you can maintain vertical boards and rebars, you are able to minimise the contortions your body would like to make, and this minimises your horse's contortions too. Finding the missing rebars gives you a blueprint for the resilient strength that is needed to ride the lateral movements well.

CHAPTER 8 NOTES

1. Myers, Thomas, *Anatomy Trains*, 3rd edn, Edinburgh, Churchill Livingstone (2014), p.133
2. Wanless, Mary, *For the Good of the Rider*, Kenilworth (1998)
3. Bennet, Dr Deb, 'Lessons from Woody', www.equinestudies.org
4. Greve, Line and Dyson, Sue, 'Body lean angle in sound dressage horses in-hand, on the lunge and ridden', *The Veterinary Journal* 217 (Nov. 2016), pp.52–7.
5. Wanless, Mary, *Ride With Your Mind Clinic*, Kenilworth (2008)

CHAPTER 9

THE DEEP FRONT LINE

An Introduction to the Rider's Deep Front Line

The term 'core' is commonly used in exercise and therapy circles to refer solely to the deep corset of muscles that wraps around the abdominal wall, encasing it from the diaphragm above to the pelvic floor below. I am presenting a much more expansive definition of core than simply high tone in the transversus abdominis. The Deep Front Line (DFL) is the literal 'core' of the body – what you would see if you cored

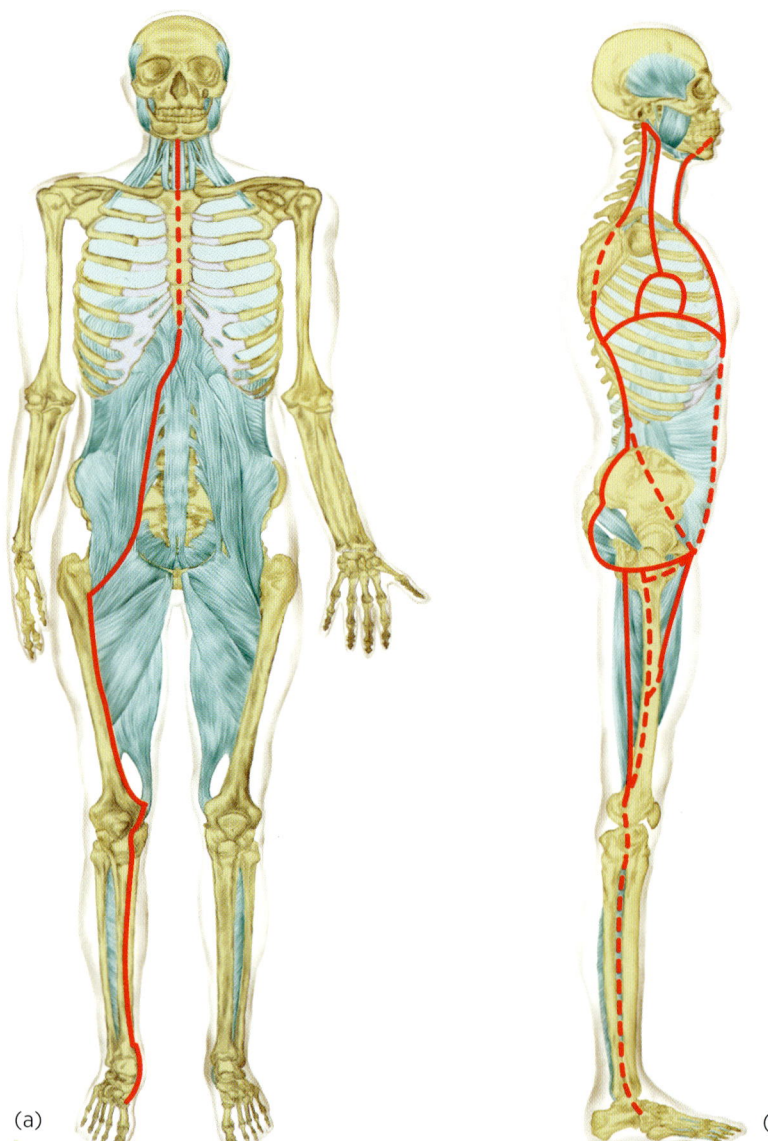

Fig. 9.1 The Deep Front Line (a) viewed from the front, and (b) from the side, showing an expanded definition of the core. This includes the deepest muscle of the calf, the adductors of the inner thigh, the psoas complex of internal muscles, the fascia of the abdominal organs, heart and lungs, and deep muscles in front of the neck vertebrae that connect into the skull.

the body from the top, like an apple. It runs from your toes up the inner seam of the legs, then on through the pelvic floor and trunk all the way to your tongue, jaw and skull. It includes the muscles that lie in front of your spine, and also the strings and sheets of fascia that weave around and through your organs, tying them to the body wall.

Here I am going to emphasise only the most important elements for riders. Even so, this chapter might seem rather 'heavy' on muscle names and anatomy. If it becomes too much, enjoy the illustrations and come back to it later. It clears some traditional confusions, and if you feel particularly challenged as a rider, it may well explain your plight and offer the antidote. It also describes the skills of exceptional riders, making them much more accessible to others.

The elite few who make riding look effortless have access to their DFL from top to toe, creating an even tension that stabilises and orients them on the horse, keeping their extremities fluidly under their control. Access to this core minimises the effects of the asymmetries and restrictions that plague so many of us. We began your journey to 'inhabit' your core on page 62 with our exercises on breathing and bearing down; I hope you are now ready to build on that foundation!

The DFL begins deep under each foot, in the tendons and muscles that make your toes curl when you get stressed (see Fig. 9.1). Since it has a branch that runs up each leg, it is simplest to describe that portion in the singular. In the foot, two long tendons that attach into your toes join another from your inner arch, and curve around your instep. They pass just behind the bony knobble on the inside of your ankle, becoming muscles that lie deep within your calf in their own fascial compartment, just behind the shin bone and its smaller partner (the tibia and fibula).

At this level, the DFL lies behind the Superficial Front Line, and in front of the Superficial Back Line (see Fig. 9.2). From the calf it passes over the back of the knee, to the bony knobble that you can feel on your thigh bone just above the inside of your knee. It then becomes both the front and back of the fascial package that contains the essential (to the rider) adductor muscles of the inside thigh.

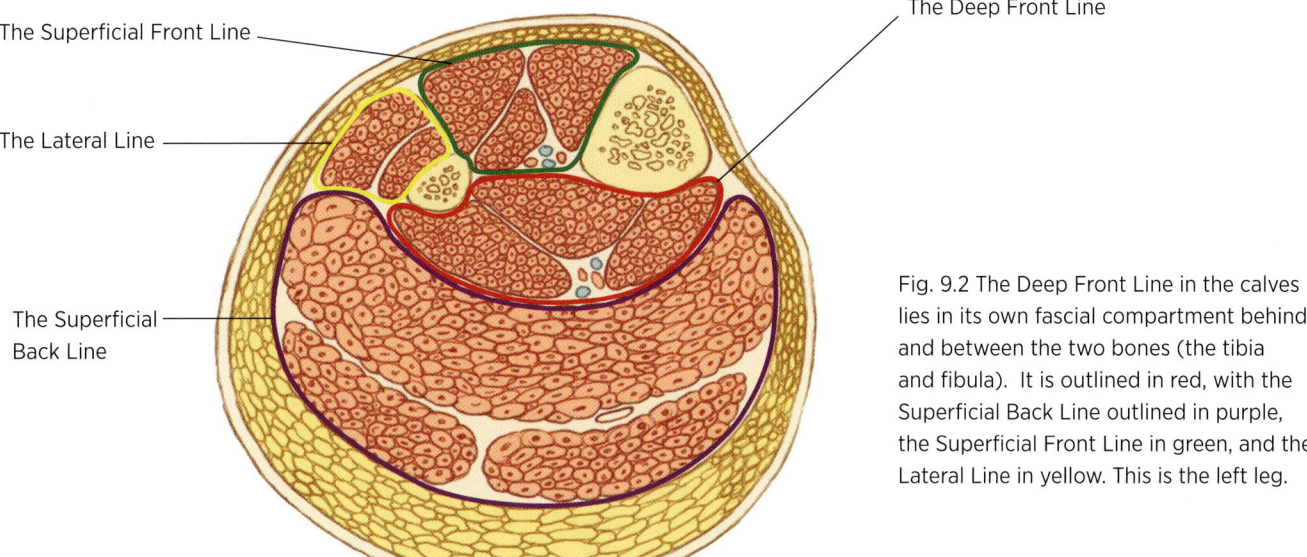

Fig. 9.2 The Deep Front Line in the calves lies in its own fascial compartment behind and between the two bones (the tibia and fibula). It is outlined in red, with the Superficial Back Line outlined in purple, the Superficial Front Line in green, and the Lateral Line in yellow. This is the left leg.

Fig. 9.3 Within your thigh there are three 'orange segments' of muscle, each in its own fascial compartment. This is the left thigh: notice how the quads make a far bigger segment than the adductors on the inside, or the hamstrings at the back. The adductors are outlined in red, to show the septa containing the anterior and posterior branches of the Deep Front Line. The black line separates the quads from the hamstrings.

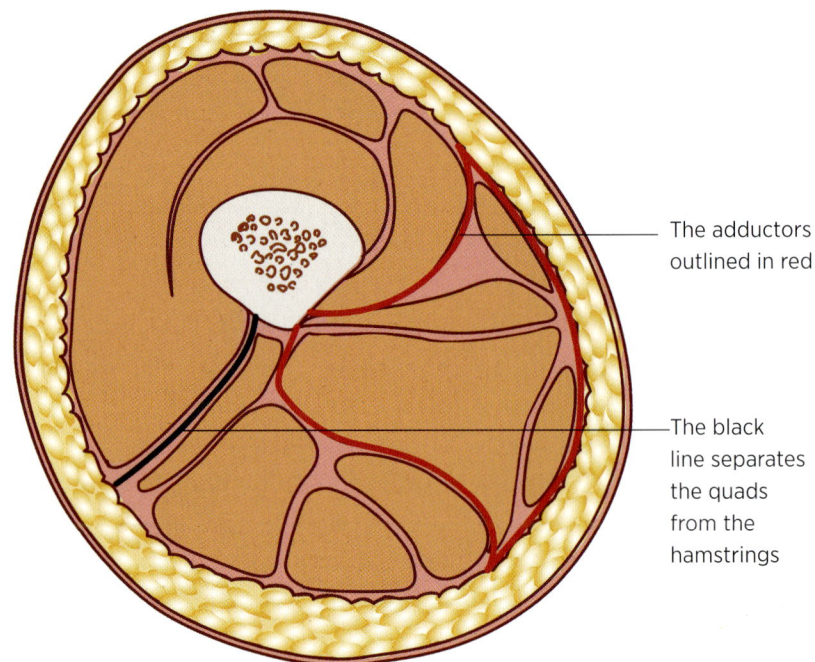

The adductors outlined in red

The black line separates the quads from the hamstrings

To understand this we need to return to our original analogy of the orange. As you eat each orange segment, you separate it from its neighbour by peeling two layers apart. Within the body, each muscle group is surrounded and separated from its neighbouring group by the equivalent two layers. These should be able to glide on each other, but in the human body they are often stuck. Together, they form a wall known as the intermuscular septum.

In your thigh there are three large septa, separating and containing three groups of muscles. (The fourth group, the abductors of the Lateral Line, lie above the thigh in the hip, and their extension, the iliotibial band, heads down towards the knee on the outside of your leg. It passes within the fascial 'unitard' that covers the 'orange segments' of muscles – in effect within the pith of the orange) (see Fig. 5.1 page 92). The three septa connect your thigh bone to the 'unitard', so imagine your thigh as an elongated orange with three segments! (see Fig. 9.3). The segment containing the quad muscles is about twice the size of the other two, covering the front of your thigh and reaching further around towards the back outside thigh than you had probably imagined. The hamstrings form the second segment at the back, and the triangle of the adductors fills up the inside.

The posterior branch of the DFL follows the septum behind the adductors, which separates them from the hamstrings at the back inside of the thigh. This branch reaches directly to the pelvic floor in several different places, connecting this line to the front of the spine. The main, anterior, branch of the DFL follows the fascial septum between the quads and adductors. The septum and these muscles connect to the straight posterior edge of the thigh bone.

The anterior branch continues with muscles that pass over the bony ridge to each side of your pubic bone, and in front of the ball of your hip joint. The most significant is the psoas (pronounced so'-as) muscle, which then dives behind your

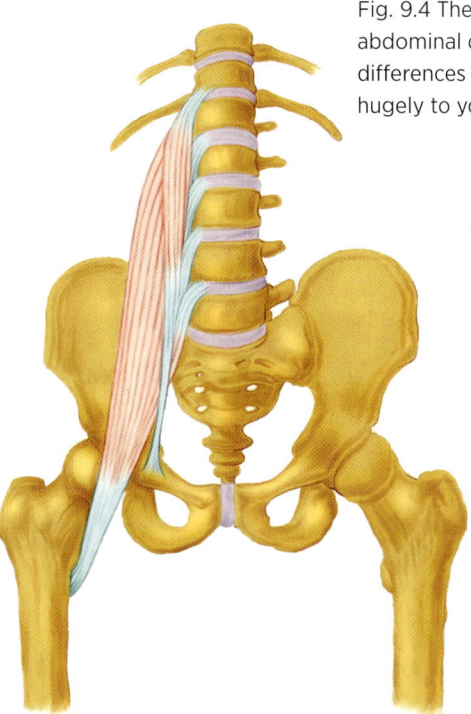

Fig. 9.4 The psoas major and minor lie within your abdominal cavity. This shows them on the right side; differences in them between right and left contribute hugely to your asymmetry.

organs on its journey from the bottom front of your abdominal cavity to the root of your diaphragm, which lies on the front of your spine (see Fig. 9.4).

The diaphragm separates your lungs from your guts and, depending on how well it works, it moves down on each in-breath and relaxes back up on each out-breath. It is shaped like the cap of a mushroom, and from the diaphragm down, we can simplify the anatomy of the DFL by thinking of a rather old-fashioned paper pin: its top surface is a dome, and its two legs would be threaded through the hole in a sheaf of papers, and bent outwards to hold the papers together. Imagine its legs separated a little, just like the right and left branches of the DFL as they extend via the psoas muscles to the inside thighs, then deep into the back of the calves and ultimately to the underside of the feet (see Fig. 9.1 page 184).

Most people think of their legs as starting at the ball and socket joint of the hip, but the leg starts *right at the bottom of the ribs*, with the psoas muscle shown in Figs. 9.4 and 9.8 page 193. Not surprisingly, the best athletes move as if their legs begin from the insertion of the psoas at the top of this inverted 'V', swinging from the bottom thoracic and first lumbar vertebrae (T12–L1) and forming much longer pendulums than if they started at the hip joint.

From the roots of the diaphragm (the crura), the DFL continues inside the ribcage. We can think of it having front, middle and back branches, although in the living body these are seamless. The front branch is the most important: it travels over the dome of the diaphragm from the back of it to the front, and then continues on upward behind the sternum. The middle track includes the fascial bags around the heart and lungs, and the posterior track runs right up the front of the spine.

The three branches all arrive at your neck in front of your neck vertebrae, and they all lie deeper in your throat than the muscles of the other lines. The DFL then ends its long journey on both your jaw and the bottom of your skull. The front branch connects into your hyoid bone, which lies at the root of your tongue, and then to your tongue itself. This means that your tongue is connected to your toes! Thus, the common tendency to anxiously curl your toes in your boots can affect your body all the way up. In fact, curling your toes adversely affects all three of the 'middle' lines – the Superficial Back and Front Lines, as well as this DFL.

This degree of connectivity has numerous implications, which we will consider as we look at regions of the DFL in more detail.

The Deep Front Line from Toes to Diaphragm

Curling Toes and Wobbly Calves

Curling your toes hugely destabilising, and many riders do this on one or both feet. Nervous riders may do it all the time, but it is most likely to happen in canter, especially when things do not go to plan! The effects travel up the DFL through

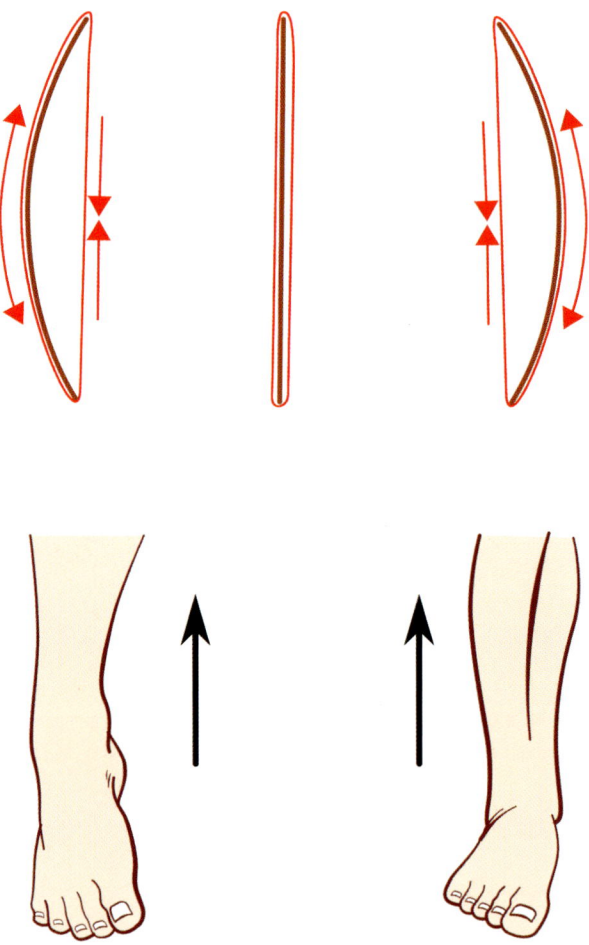

Fig. 9.5 The Deep Front Line and Lateral Line work like two strings, one on each side of a wooden bow. When one string is pulled up, this contributes to 'X' legs, 'O' legs, and contorted feet.

the deep muscles of the calves, into the inside thighs, and perhaps even higher. To counteract the tendency, lift and spread your toes. This makes the tendons around your ankle stand out (see exercise page 81).

Curling your toes tends to go with curling your fingers, and is part of a prehensile grasping reaction that occurs within the older parts of the brain. This might have helped you to grasp a tree branch before you fell into the jaws of a hungry tiger, but grasping the stirrups with your toes will not save you! If you hold the reins without bending the last joint of your fingers you are less likely to curl your toes, and this is reason enough to advocate 'pads against palms' (see Fig. 7.8 page 143).

Feet that become contorted when you ride are even more common than toes that curl, and we have already seen how the Spiral Lines can distort the feet. In addition, think of the DFLs and the Lateral Lines like bowstrings on each side of a wooden bow (the bones of your thigh and calf – see Fig. 9.5). Your legs and feet are affected by the imbalances between the lines: if the Lateral Lines are locked short and the DFLs locked long, you will be knock-kneed ('X legs' or genu valgus). Your inside thighs will tend to be tight on the saddle, and you may grip with your knees. The outsides of the feet are pulled up, and your weight is taken towards your big toes, so that an observer can see the sole of your boot when you ride (feet pronated). The opposite scenario yields bow legs ('O legs' or genu varus), with weight taken towards your little toes (feet supinated) and your knees coming off the saddle.

For many years, sticking out the tendons around the ankles was the best fix I could offer for contorted feet and wobbly lower legs, and it worked to a degree (see exercise on page 81). Adding the stability offered by the Functional, Lateral and Spiral Lines helped significantly, as enhanced by the resistances of Figs. 3.16(a) page 76, 3.21 page 80, and 5.17 page 117. But if we can access the DFL within the calf, we change the game! The front of its compartment is formed by a membrane that joins your tibia and the smaller fibula. The latter begins below your knee as a bony knobble towards the back and outside of the top of your riding boot, and it forms the outer part of your ankle (see Figs. 9.1, 9.2 pages 184-185).

To get the DFL properly engaged:

1. Sit on a firm chair, and push your ankles out against the chair legs. This engages the Lateral Lines.
2. Repeat this, but now imagine the membrane, and how your calf would look in cross-section (see Fig. 9.2 page 185). Can you think of tensioning the membrane, as if you could spread it from the inside to the outside, ironing it out in that direction, and involving it in this push?
3. How much difference does this make? Did you naturally involve the membrane the first time you did the resistance? In my experience, only elite riders have this 'wired in'.
4. Especially when you tension the membrane, you might find that your toes automatically go up and spread, and that the tendons around your ankle and knee stick out. This shows that you are also engaging the Superficial Front and Back Lines.

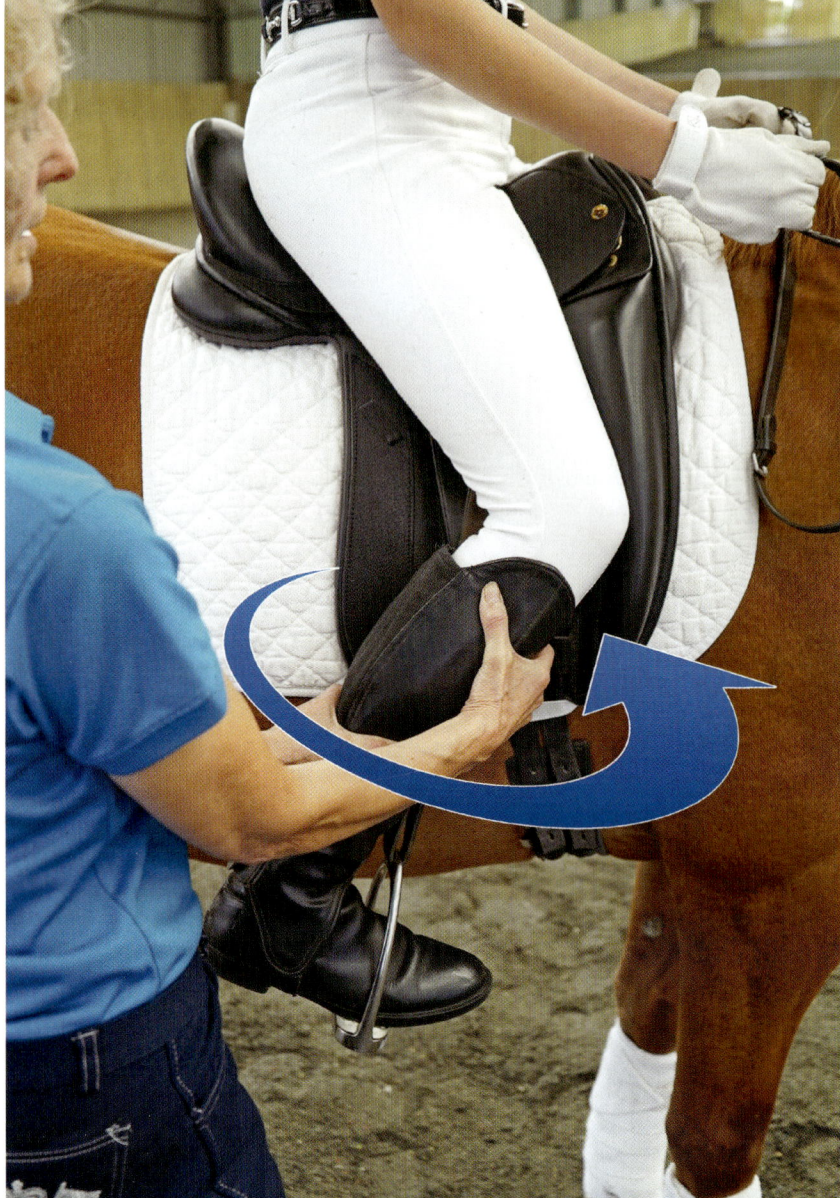

Fig. 9.6 'Fibs round tibs': rotating the calf so that the outside of it comes towards the front and the toes go more in.

5. As a variation on the theme, and especially if you tend to turn your toes out, think of rotating the fibula around the tibia. In coaching, we help the rider by putting our hands around the calf and twisting it so that the outside moves towards the front, and the inside towards the back (Fig. 9.6). We call this 'fibs round tibs'.

6. For a complete calf workout, push your ankles out against a resistance whilst tensioning the membrane, and lifting your toes as you stick out the tendons around your ankle and behind your knee. Remember to breathe! Do you also need to think of 'fibs round tibs'?

7. With all of the above in place, put your heels outside the chair legs, and kick them a few times, using a quick movement that rebounds. Think of your leg aid as a slap and do not use more strength than this. You are not trying to *make* your horse move: your aim is to teach him the meaning of this signal, and to make sure that he takes it seriously.

The Thighs

The balance between the inside and outside thighs

A dynamic tension between the DFL and Lateral Line yields a thigh that lies on the saddle 'like a damp dishcloth' – a traditional image that suggests a muscle quality that few riders achieve. Along with this ideal quality comes a stable lower leg and a foot that is evenly weighted from big toe to little toe. I have already introduced the inside thigh muscles on pages 73-81, in Chapter 3 (Superficial Back and Front Lines) and pages 105-117 in Chapter 5 (Lateral Lines): the vast majority of riders need this preparatory work on them, and we now consider the dynamic balance between the inside and the outside thigh.

The 'how' of this has never been well taught, because instructors of all generations have lacked a clear understanding of how the thigh muscles work in riding. If you are an older rider, you probably grew up being told to 'grip with your knees'. This suggests that the contact at the knee should be stronger than the contact at the top inside thigh, and is the fatal flaw in this advice. If you grip with your knees you may well 'ping' up and back off the horse, like an old-fashioned clothes peg. Also, the pressure of your knees might stop your horse from wanting to move freely.

In the early 1970s, along came the antidote to this, as riders were told to 'relax your thighs and take your knees off the saddle'. The was driven by the fear of 'clothes pegs', and the realisation that the many muscles of the inside can easily overpower the few on the outside. It has been the doctrine ever since – and whilst it may stop riders from gripping, it is not a good description in language of how skilled riders use their outside thighs.

Good use of both the DFL and Lateral Line within the thighs invokes a surprising result that contains an inherent contradiction (and a damp dishcloth!). To get proper balance between them, your inner thighs must first lie on the saddle, so that they support your bodyweight and are part of your sitting surface. Remember the repositioning of Fig 3.6(b) page 59.

1. Make sure that your thighs are well rotated in. Then use the resistance of your fist and/or fingers on the pommel (see Figs. 5.9, 5.10, 5.11 pages 105-106) to help you get your top inside thighs to lie snuggly against the saddle. Inevitably, one side will need these resistances more than the other.
2. When you can easily maintain this, imagine that your inside thighs are covered with suction cups that can draw the horse's ribcage *outwards*, encouraging fuller breathing, and a more fluid step.
3. Realise, however, that you can only use suction on a surface *after you have made contact with it*. So having your thighs *off* the saddle makes suction impossible. It does not work to think of 'making a space for the horse to fill in to' as you hold one or both thighs off the saddle.
4. If your attempt to make suction caused your thighs to lose contact with the saddle, you need to narrow in, make contact, and start all over again.
5. Thus, in a paradoxical way, the thighs narrow in and widen out *both at the same time*, bringing both the inner and the outer thigh muscles (the DFL and the Lateral Line) into play.

Riders who do this well have little idea of what they are actually doing, and this is one of the areas where a skilled rider's self-report is least likely to be an accurate description of her muscle use. This floors riders lower down the totem pole. For instance, some elite riders who have an exceptionally firm contact with the inside thighs will tell you that they are 'doing nothing', which simply means 'nothing that I register as unusual'. The muscle use I am describing here is a very big 'something' for most riders, and you will not find it by trying to emulate a 'big name's' notion of 'nothing'!

The three orange segments of the thigh

Within the thigh, the involvement of the two septa that contain the branches of the DFL means that we are accessing the thigh from skin to bone. This gives a sense of *substance* (known as 'hydraulic amplification' in the biomechanics world) that yields a lot more stability than just focusing on its surface (see Fig. 9.3 page 186). We will also extend our thoughts beyond the DFL, to include the third septum that lies on the back outside of the thigh, between the outside of the hamstrings and the back of the quadriceps group.

This exercise can be done in a chair, and then whilst riding in walk.

1. As you sit in your chair, mentally rehearse riding in walk. Think of the inside thighs rotated in on the saddle and bearing weight, so that you are 'kneeling'. Remember how it feels when your legs are repositioned as in Fig. 3.6 page 59, and also after the resistances of Figs. 5.9. 5.10, 5.11 pages 105-106.
2. With the analogy of the elongated orange in mind, think of the three septa that contain the three compartments of the thigh, appreciating that each one goes all the way from skin to bone. Then notice the three compartments of your thigh. Which ones are most and least 'filled in' in your awareness?
3. If your thighs were a three-dimensional painting, would each compartment be filled in with equally deep colour? Would parts of them be a pale watercolour, and parts of them deep acrylic colour? Can you deepen the watercolour, making it more like the acrylic colour and even this out? Alternatively, imagine yourself as a stuffed toy rider. Which compartments would have most and least stuffing?
4. Is the pattern the same or different within each thigh? Does the filled-in-ness differ higher up and lower down on the thighs? How clearly can you imagine the septa?

Fig. 9.7 Straighten your legs and cross your ankles, resisting them against each other. Do it both ways, and notice where you firm up in each case.

5. As you sit in your chair, straighten your legs and cross your ankles, resisting them against each other (see Fig. 9.7). Does this help you to find the septum at the back outside thigh, and to 'fill in' the outside segment?
6. In the ultimate of 'kneeling' and spreading weight down through your thigh, you would kneel down the length of the inner posterior septum, from just in front of your seat bone to the back inside of your knee.
7. Think also of the septum in the back outer thigh, imagining it as a firm structure that sticks out from your thigh bone, reaching to your skin. Can you re-create the clarity you get when you resist your crossed your ankles against each other? This can make a huge difference to your ability to access your outer thighs, helping to keep you more straight and stable.

'Leg-pits'

Before we leave the inside thighs, we have a huge amount to gain from considering the top part of the septa at the front and back of the thigh. This is the area of the 'femoral triangle' which Myers calls the 'leg-pit', as its configuration is so similar to the armpit.

1. Reach across your chest with one hand, and put your fingers inside the front of your armpit, with your thumb on the outer surface. In the ridge between your thumb and fingers lies a strong tendon, which firms up if you pull down your shoulder and elbow (see Fig. 6.2 page 124).
2. Your thumb and fingers lie on two surfaces that can slide across each other: you can slide your fingers deeper into your armpit, as you slide your thumb away from your shoulder girdle. You can also reverse this, sliding your thumb towards your collarbone as your fingers slide slightly out of your armpit.
3. Your leg-pit has the same structure. Between each surface is the strong and obvious tendon which inserts the adductor longus into your pubic bone.
4. Take this tendon in your hand, placing your thumb on the top and your fingers underneath. (Don't be shy, it's only flesh!) Much of the upper surface beneath your thumb is formed by the pectineus muscle, which is an important part of the DFL (see Fig. 9.8).
5. Just as before, your fingers can slide further underneath you as your thumb slides away from your groin, or you can also reverse this movement.

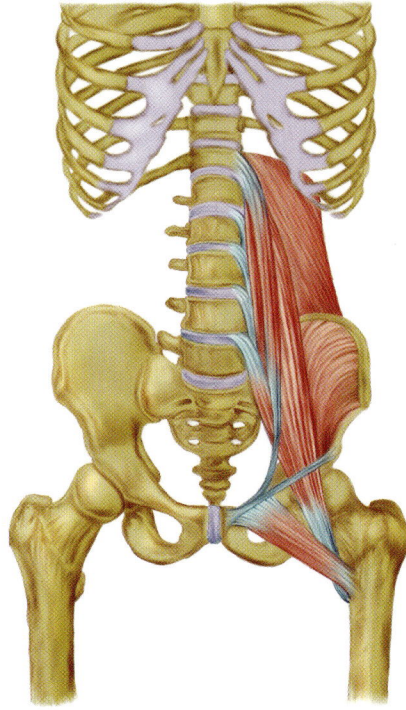

Fig. 9.8 The upper surface of the leg-pit is formed by the pectineus and psoas muscles. The drawing also shows the left iliopsoas complex in its entirety, including the two tracks of 'locals' described later. The quadratus lumborum and iliacus muscles form the wider of the two 'locals', and the other set is formed by the pectineus and psoas minor. The cross-section of Fig. 5.13 page 109, Lateral Lines, also shows the psoas muscles.

6. If you are a rider who has forever tended to tip forward (regardless of how many times you have been told not to!) you are about to discover the antidote to your problem. *You tip forward because the underneath surface of your leg-pit keeps sliding backwards relative to the top surface.* It disappears back under you, closing the angle between your thigh and torso.
7. Think of keeping your 'leg-pits out in front of you', so that their underneath surface would keep sliding *forward* relative to their top surface. You will suddenly find that you no longer lean forward (as long as you remember to keep doing it!).

Think of these septa as guy-ropes. At the front inside thigh is a guy-rope attaching your pubic bone to the front inside of your knee. The septum at the back inside thigh is a guy-rope attaching your seat bone to the back inside of your knee (see Fig. 9.9). If you cannot stop yourself from tipping forward and hollowing your back, the front one is too short, and the back one is too long. In addition, the fascial layers within the septa could be stuck together, and this creates restrictions that affect the position of the pelvis, fixing it in the tilt of a round or hollow back.

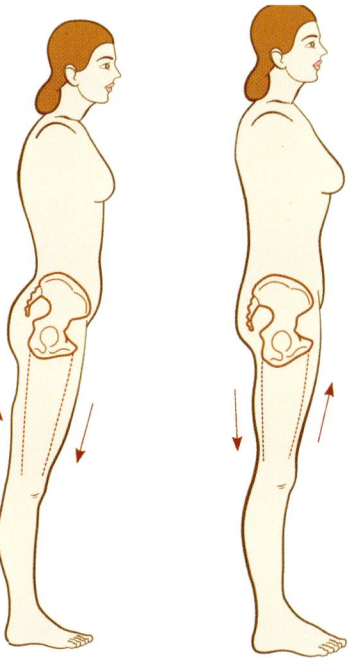

Fig. 9.9 The septa act as guy-ropes that can restrict the ability of the pelvis to move between flexion and extension, potentially leaving you stuck in one of those positions. The person on the left has the pelvis pulled down at the front and an extremely hollow back. The person on the right has the pelvis pulled down at the back.

Keeping your leg-pits out in front of you evens out the guy-ropes, and greatly increases your ability to resist sliding back down the 'mantrap' of a horse (see page 86) with a strong push back. Otherwise, when he hollows his back, the underneath of the leg-pit (which is already too long) adapts to the new shape of the (disappearing) surface. Keeping your leg-pits out in front of you has proved a remarkably quick and robust fix for what had seemed, for many riders, to be an intractable problem.

Conversely, if you are a rider whose shoulders are thrown back on the third beat of the canter stride, think of sliding the underneath of your leg-pits *back underneath you*. This will help you to keep your shoulders and backside still throughout each stride, with your weight resting on the ideal part of your underneath.

The Pelvic Floor

Our next port of call is the sitting surface of your underneath. Your entire pelvic floor is a diamond shape: the points at the front and back of it are the pubic bone and the tail bone (coccyx). The points at the sides are the seat bones. The front two sides of the diamond are formed by the bony rami which join your pubic bone to your seat bone on each side (see Fig. 9.10(a)).

The rami are like the rockers on a rocking chair or the runners on a sledge, except that as well as sloping forward and up they narrow in towards each other fitting (if you love your saddle) within the seam lines on the seat, and (if you love your saddle) matching the slope up towards the pommel (see Fig. 9.10(b)). The back two edges of the diamond are formed by strong sacrotuberous ligaments that join the seat bones to the tail bone.

Fig. 9.10(a) The pelvis placed on the saddle shows the diamond of the pelvic floor, with the bony rami joining the pubic bone to the seat bones, and strong ligaments joining the seat bones to the tail bone.

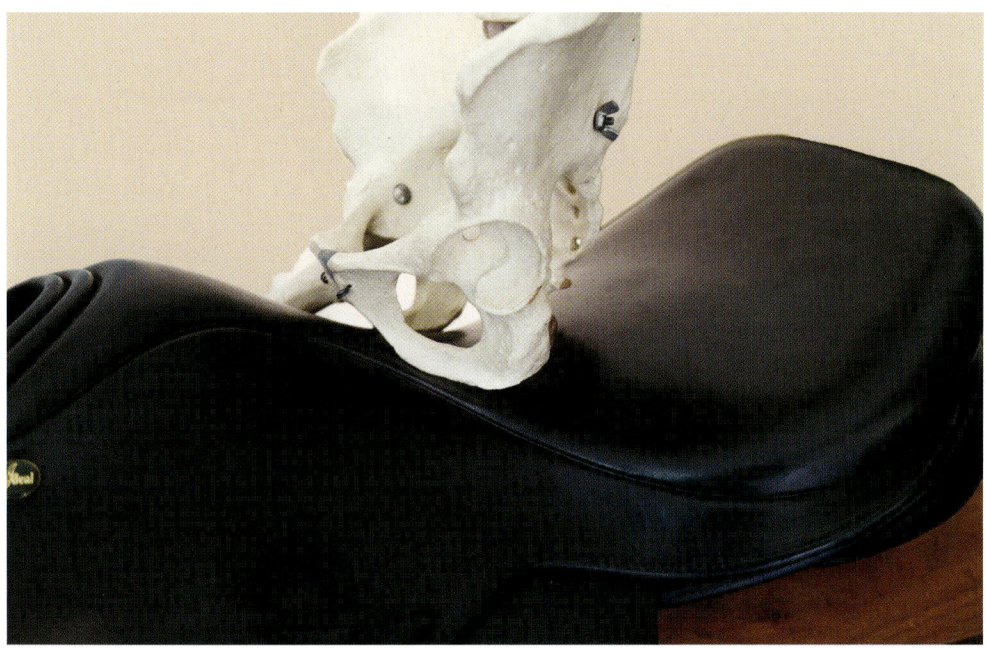

(b) This shows how the rami which join the pubic bone to the seat bones fit neatly (if you love your saddle) within the seam lines of the seat.

Most riders are familiar with their seat bones, but few realise that the pointed bones they feel continue forward as ridges. This makes a nonsense of the idea of a 'three-point contact', where the third point is the pubic bone. To add to the confusion, most riders know approximately where their coccyx is, and some (especially men) mistakenly imagine that *this* is the third point.

However, if your coccyx were to become a pressure point in your sitting, you would be in pain, as well as being way too tucked under. Also, some men think that their lesser trochanters (another pair of bony knobbles that we will meet soon) are their seat bones. I would not vouch for the accuracy of our traditional literature, and the potential for confusion is huge!

In most cases your weight should be taken on the *front half* of the diamond of the pelvic floor, since the back half of it is too far behind the horse's centre of gravity and the strongest part of his back. It needs to have *contact* with the saddle, but to bear very little *weight*. This conundrum is challenging for many riders: hollow-backed riders do not easily get the back triangle in contact, whilst round-backed riders easily put too much weight on it. Meanwhile, an observer can help both groups of riders by checking that the underneath of their torso-box does not distort: it needs to maintain the correct distance between the vertical back and front.

Most riders are familiar with the adductor longus (and its obvious tendon), which inserts at the corner of the pubic bone. But few realise that more adductor muscles insert *all along the length* of the rami, with only the posterior part – the seat bones – reserved for the attachment of the hamstring muscles (see Fig. 9.11). Thus the adductors form a significant 'orange segment', which is wrapped in the back by the posterior muscular septum.

The septum continues onward and upward past the adductors and the inside edge of each ramus, connecting into the muscles of the pelvic floor. This is a sling

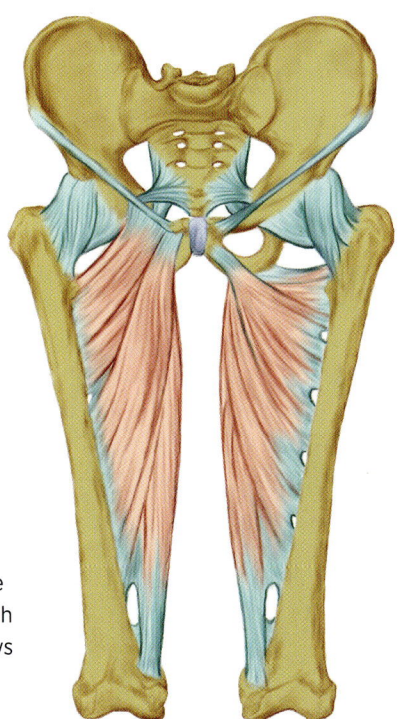

Fig. 9.11 Various adductor muscles insert along the length of the rami. The hamstring muscles (which you cannot see here) attach into the seat bones at the very back of them. The left side shows the muscles on the surface of the leg-pit. These have been removed on the right side.

– a funnel hung within the bowl of the pelvis – designed to support your internal organs.

1. Would you think of your pelvic floor from the front to the back, or from the back to the front?
2. Hollow-backed riders, with disappearing leg-pits and an underneath that is too long, choose the first option. They have an overly slack pelvic floor which is too 'down-loose'.
3. Round-backed riders would think of their pelvic floor from the back to the front, and so would most well-organised riders who keep neutral spine.
4. Do you constantly pull your tail bone towards your pubic bone as you ride? This is a helpful idea for some hollow-backed riders, but for many others it would make the pelvic floor *shorter* than its rightful length as the bottom of the torso-box.
5. Pull your tail bone towards your pubic bone: notice how this shortens the underneath of your torso-box, whilst encouraging you to round your back. You then topple back behind the ideal balance point, putting too much weight on the back triangle. This strategy is more common in men, whose pelvic floor is stronger by virtue of having one less orifice; but I see it too in women.
6. When you next ride, have an observer give you feedback about the length of the underneath of your torso-box and the verticality of the box. Is the underneath too long, too short, or just right?
7. Finding 'just right' in your pelvic floor can make a significant difference, and the Goldilocks Principle applies here!

If I watch a rider and think 'her breeches are too small; they are stopping her underneath from coming down on to the saddle' I give her the benefit of the doubt. But by the time I have thought this over several lessons, seeing several pairs of breeches, I am left wondering when and how to break the news! She is holding her pelvic floor too much up and too tight: usually, I quietly suggest that she makes sure that she rides as if she 'could pee and could pooh'.

If you ride doing a Kegel (contracting to stop yourself from peeing), it gets you into really big trouble by making the pelvic floor too 'uptight' and unresponsive to the changing forces of the horse's gait. Try this sitting in your chair:

1. Make a Kegel. You may need to exaggerate it to realise that it lifts your pubic bone, and makes you suck in your bikini line (which some men have called the 'speedo line'). On a saddle this would be inevitable.
2. A strong Kegel also makes you roll back into a round-backed position with your weight more on the back triangle of your underneath. Your thighs then rotate outwards and come off the saddle. Do the exaggerated version, and realise that you are moving beyond a true Kegel by scrunching your deep butt muscles. This imitates true tone in your pelvic floor, but is not a good substitute!
3. If you were attempting to stop or turn, you would have no bear down and no inside thighs to help you. All you have left is pulling on the reins! This alone is

bad news – but I also suspect that it is the Kegel that creates in 'an electric butt'. So when the tension of the Kegel makes your horse hyper, this worsens your plight!

4. You have expended a lot of muscle power, and might mistakenly believe that you are bearing down. To get from the Kegel to bearing down, you would have to drop your pubic bone and push against your bikini line, bringing your weight onto the front triangle and your thighs onto the saddle. Clear your throat as you do this.
5. Re-learning how to bear down can be a long process, and every time you feel stressed you will almost certainly make a Kegel by mistake. Do you curl your toes as well?
6. Some of the ideas from Pilates can be confusing here. If anyone suggests that you need to draw up your insides, you probably do *not* need to do that whilst riding. Think of dropping your underneath down onto the saddle.

Pelvic floors are a big deal for riders. They can also be very different to the right and left of our mid-line, contributing significantly to our asymmetry. We will come back to this when we think about the diaphragm.

The Psoas Muscle – the Core in the Torso

From the leg-pit, the anterior branch of the DFL continues as the psoas major muscle. Whilst many muscles connect your legs to your pelvis, the psoas muscles have the remarkable privilege of being *the only muscle that connects each leg to your spine*. This makes them tremendously significant, yet few riders know of their existence (see Fig. 9.4 page 187).

The psoas inserts into the lesser trochanter – a small bony knobble that lies on the underside of the greater trochanter, which you found in Fig. 5.2 page 93. It is these knobbles that men may mistake for their seat bones. (The shape of the female pelvis makes this much less likely.) The pectineus muscle that forms the front of your leg-pit inserts here too (see Fig. 9.8 page 193).

From the lesser trochanter of each femur, the psoas muscles pass over the front of the pelvis to each side of the pubic bone, and then over the front of your hip joints. They climb through your abdominal cavity, passing behind your guts and inserting into the sides of each of the lumbar vertebrae, progressing to their highest insertion, either on your first lumbar vertebra or on the vertebra of your lowest rib.

The psoas muscles form an upside down 'V' (remember the analogy to the paper pin), and each successive insertion from each and every lumbar vertebra adds to their substance as they travel down towards your legs. In the point of the 'V', the psoas blends with the attachments of the diaphragm, so the psoas muscle 'relates the wave of breathing to the rhythm of walking'.[1]

The psoas is the *filet mignon* in cattle, and the tenderloin in pigs; it is tender because it is composed of slow-twitch endurance fibres, and no fat. Remarkably, anatomists still argue about its functions, with some thinking that it rotates the hip inwards, and others outwards. Myers argues it does neither, and that the upper

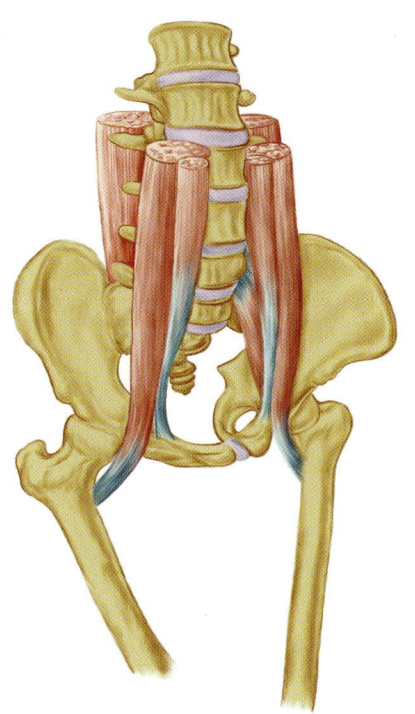

Fig. 9.12 The psoas muscles and the long back muscles fit into gullies on each side of the vertebrae, and when the tension within them is balanced your spine is stable. This is visible too in the cross-section of Fig. 5.13 page 109, in the Lateral Lines, Chapter 5.

part flexes the lumbar spine (which would round your back) whilst the lower part extends it (which would hollow your back). This suggests that all of us – and riders in particular – can stabilise the lumbar spine in the neutral position between flexion and extension by balancing how we use both ends of the psoas. It is just (just!) a case of learning how to do this – realising that it could, of course, be significantly harder on one side than the other.

Recent research implies that toning the abdominals has much less effect on reducing back pain than has been previously thought.[2] Instead, the psoas muscles are the key, and they need to have a dynamic balance with the long back muscles which are part of the Superficial Back Line. Each of these lies in a gully behind and to each side of the vertebrae; the psoas muscles lie in the gullies in front and to each side. If these four lines of pull are all well balanced, you will have a stable, pain-free lower back, and every chance of sitting the trot well (see Fig. 9.12).

Functional and compromised psoas muscles

The two psoas muscles will never be completely equal because of postural deviations that, at a minimum, involve handedness and footedness, and perhaps because of injury. Their path through the abdominal cavity is reminiscent of the boards from Chapter 5, and thinking of 'both boards on' helps you to find and equalise them (see page107). When one is really strong, the other can all but give up its role, and you might feel as if there is little more than mush in that side of your torso!

Skilled riders get really close to finding functional symmetry within the right and left DFLs, homing in on this through a myriad of tiny breakthroughs throughout their riding lives. However, many of us 'freeze' ourselves from the pelvis upwards, and we may 'freeze' our diaphragms as well. This leaves us walking only from our hip joints, and with very little breathing capacity.

Suppose your coach wants you to get the hollow out of your back, but when you think about this, your knees comes up on the saddle. Then when you think about your knees reaching lower down, your back hollows. And so it goes on … with one end of the psoas muscle reaching away from the other, as that end declares 'But I can't accommodate this extra length!'

Also, restrictions in the DFLs around the hip joints and thighs can *stop differentiated movement between the pelvis and leg* on one side or both, pulling us into hip flexion. Almost all of these habit patterns are traceable to the body's natural fright reactions, and sadly, fear, anxiety, and other emotional issues sometimes stay with us (as mindset and a 'freezing' of our tissues) for life. As mentioned earlier, toes that curl are another expression of this vulnerability.

The Two Tracks of 'Locals'

Help is at hand! Enter the two sets of 'locals' that follow the same track as the psoas major. This means that *three* separate tracks of muscle and fascia make their way from your inner hip to the back of your diaphragm. Together they form what is known as the iliopsoas complex (see Fig. 9.8 page 193). When a rider's hollow back does not easily respond to equalising the tone in the Superficial Front and Back Lines, the DFL is often the culprit, pulling the lumbar spine down and forwards towards the front of the pelvis.

The first set of 'locals' includes the pectineus muscle, which forms the upper surface of your leg-pit. At your pubic crest it is continuous with the psoas minor, which lies in front of the much bigger psoas major and also closer to your mid-line. Together, these 'local' muscles cover the same territory as the psoas major (an 'express' muscle)– see Fig. 9.8 page 193.

1. Sit in neutral pine, and push slightly back off your feet. Imagine your leg-pits connecting into the back of your diaphragm via the psoas minor muscles, which travel diagonally back and up through your abdomen, forming an inverted 'V'.
2. Can you get any sense of this connection inside you? Is one side more difficult than the other? Are both sides 'mush'?
3. Quite often, riders can get a sense of some substance at the bottom front of the abdomen, and also at the top and back, but these areas are not joined up. Remember the idea of the paper pin. Can you get those areas to join up on each side?

The second track of 'locals' also adds enormously to your ability to access the strength of your core. It lies to the outside of the psoas major (see Fig. 9.8 page 193). The lower muscle is the iliacus which, like the psoas major, inserts into the lesser trochanter. It extends upwards to cover the *inside* of the bony wing of the pelvis. Yes, there really is a muscle here! If you put your hands on your hips, they are resting on top of those wings, which I think of as the filling in a 'bone sandwich'. The 'bread' is the iliacus muscle on the inside, and the gluteal muscles of the Lateral Line on the outside.

When both 'pieces of bread' are firm rather than 'soggy' the bone itself feels strong and stabilised. When the muscles are weak, they prevent force being transmitted from the back to the front via the sides of the pelvis, so the rider feels more 'limp' and disorganised. The bread in one sandwich is often much firmer (dark rye bread compared to 'cotton wool' bread!) and this contributes to the tendency to lean and rotate inwards on a circle.

1. Put your hands on your hips to feel the bony rim of the pelvis, and then place your thumbs inside the rim to feel the iliacus working in there – lift your knee a little to make it 'pop'.
2. Push your heels outwards, against chair legs, or against a friend's resistance when riding. Can you feel the outer muscles firm up?
3. Next, think of bringing the front points of your hips towards each other, as if they were cross-eyed. Can you feel the iliacus muscles firm up on the inside?
4. Can you do both of these at the same time? Do you have the same strength on both sides?
5. Practise until you can access both the inner and outer muscles. When you can, you will also be bearing down strongly, and you might be shocked by the tone and strength you have generated. Make sure you keep breathing!

The Quadratus Lumborum (QL)

Continuing upwards, the iliacus connects to the quadratus lumborum, or 'square of the back', also known as the QL. These muscles lie on each side of your spine, filling in the space between the back bony rim of your pelvis and your bottom rib. Pilates teachers regard them as part of the 'inner corset' of muscles around your abdomen – they are the deepest muscles of your back, lying right behind your guts, and their deepest surface connects to the iliacus. If you put your fists each side of your lumbar spine with your knuckles facing each other, you are laying them on the muscles that overlay the QL (see Fig. 9.13). Tip from side to side to feel these muscles tighten on the side you are leaning away from.

Fig. 9.13 The quadratus lumborum (QL) lies under the rider's hands, and is the deepest layer of muscle behind the guts (see also Fig. 5.13 page 109).

The quadratus lumborum sits on top of the bones of the pelvis, resting like a square panel on top of a (bony) fence. The Superficial Back Line lies behind it, and the psoas in front of it (see Fig. 5.13 page 109).

1. You accessed the QL in the exercise on page 73 where you pushed on the back of a chair or the saddle. Revisit this now. Have you been able to take this strength into your riding?
2. Think of the QL sitting on the top of a fence. If you are riding along and you feel as if it falls off the outside of the fence, you are left accessing only the Superficial Back Line.
3. Sit in a hard chair with both seat bones pointing straight down, and very subtly push back from your feet. Think of extending the front triangle of your pelvic floor back, elongating it as you search for the muscle that forms its back edge and connects your seat bones. We will call this 'the crossways muscle' rather than 'superficial transverse perineal'!
4. Does this make the muscle clear to your felt-sense? Imagine it if not. Take time with this, first on a chair and later in the saddle, to progressively make it clearer.
5. Can you stack your QL vertically over this muscle? This may make you feel slightly hollow-backed – sit sideways on to a large mirror to check this out. Remember that neutral spine is the place where your spinal curves are balanced, not the place where your entire spine is straight.
6. You have placed the QL directly over the furthest back line in your underneath that gives you a really firm surface to sit on.
7. In sitting trot, the stability that you gain when the QL is stacked over the 'crossways' muscle is huge. I have seen it add enormously to the overall strength of many riders, improving their ability to stay 'with' the horse in both trot and canter, and to push their hands forward.

Intractable issues of mid-back and hips

If problems around your mid-back and hip joints remain intractable, your best help lies in skilled bodywork or rehabilitation exercise that encourages the psoas and the diaphragm to 'let go'. These core muscles rarely need resistance training, which might benefit your pecs or lats – it is balance left and right, even tone through the muscle, and coordinated participation that make the difference in riding, or any other sport.

The results could be life-changing, and might even spare you from needing more aggressive measures like a hip replacement in later life! Careful stretching can help too; the warrior poses in yoga (lunges in athletic training) address the psoas muscles; but to be effective, stretching needs to be done with a level of precision and attention that few people achieve. Given the depth and obscurity of the DFL, and the fact that outer muscles can substitute for most of its functions (other than breathing), it is harder to improve the functionality of the DFL than any of the other lines, but the benefits are enormous.

The Diaphragm as Two Squares

The psoas major and minor, and also the QL, all intertwine with the back of the diaphragm, which is one of the strongest muscles in the body. Remember the idea of the paper pin, with the diaphragm as its head, and the psoas muscles as its legs. If the head of the fastener is very asymmetrical, this affects the whole of your torso-box. That asymmetry would almost certainly be mirrored in how your pelvis meets your horse!

In this exercise we are going to think of the diaphragm as a flat surface, even though it is, in reality, a dome.

1. Think of your torso like a three-dimensional capital H, with a vertical front and back, and with the diaphragm as the horizontal connection that joins them.
2. When you have 'both boards on' (see page 107), imagine your diaphragm in each outer third as a square (see Fig. 9.14). The diaphragm has a central tendon that runs from the back of it to the front of it, so think of this filling the (much narrower) middle third of your torso.
3. Imagine each outer diaphragm-square as a rebounder (a small trampoline). In reality your diaphragm is connected all around your ribcage, but imagine how the most important springs attach the rebounders at each corner – to the front and back of your boards, and the front and back corners of your torso-box (see Fig. 9.14).

Fig. 9.14 The diaphragm squares are like two rebounders, attached at each corner by strong springs. Between the squares is the smaller middle third.

4. Can you imagine each of these springs? Are some of them not attached, making parts of the diaphragm soggy and less active? Imagine stretching the rebounder as you hook each spring to the frame. How many springs need reconnecting? Where are the connections most clear?
5. Do you have 'diaphragm droop', where the diaphragm is not well attached at both front and back on one side? If this did happen on one side, which side would it be? Can you somehow level the diaphragm up and attach those springs? Your pelvic floor may 'droop' in the same way – does your horse 'droop' that way too?
6. If you can level out those two surfaces in yourself, you will almost certainly level him out as well. Reach up with both arms and breathe in deeply – you will be able to feel where the diaphragm is not connected.
7. Think of the surfaces of your diaphragm and pelvic floor as the wings of a biplane, and your middle third as the body of the plane. Can you level up your 'wings'?

The most subtle way that riders fall down the 'mantrap' (see page 86) goes beyond distortions in the Superficial Back and Front Lines, and results from the back of the diaphragm being 'droopy'. The soggier the horse is behind the saddle, the less support there is for the diaphragm, QL, and your entire back. Unless you can use your 'well-strungness' to 'restring' your horse, everything behind the crossways muscle will probably 'droop'.

1. Is the back of the diaphragm 'hooked on' and level with the front of your diaphragm, or is it lower and soggier? Is it less 'there' to your felt-sense?
2. Can you make your diaphragm feel more firm and level? You may well have to build into this over time. Start by attempting to get a sense of a horizontal connection that begins in your front behind your lower ribs, and extends back through you.
3. Can you gradually extend this further back, and have it meet the horizontal surface that begins from the back of your lower ribs?
4. As you become more firm and level in your diaphragm, you have every chance of changing your horse's back, so that it too becomes more level, firmer, and more 'stuffed' both under and behind the saddle. This will make it easier to keep the QL stacked over the 'crossways' muscle that joins your seat bones.

The Deep Front Line above the Diaphragm

We left the posterior muscular septum of your thigh at the point where it attaches into your rami, pelvic floor and coccyx. From there it continues upwards via a ligament that runs up the front of your spine (the anterior longitudinal ligament), passing behind your lungs and heart to your neck (see Fig. 9.1 page 184). This branch of the DFL, the furthest back of three, also includes muscles that lie to each side of your neck vertebrae. They connect into your atlas, the first vertebra of your neck.

We will talk more about this pair of muscles called longus colli (long muscle of the neck) when we consider how we as riders can influence the horse's DFL.

The upper *middle* track of the DFL is a continuation of the psoas muscles, which intertwine with the QL and the back of the diaphragm, taking the line of pull into the central tendon of the diaphragm. This goes from the spine to the highest peak of the domes and connects to the membrane that joins your sternum to your spine, dividing your ribcage and separating your right and left lungs.

The pericardial sac (containing the heart) lies within this membrane and rests on the central tendon, so effectively, your middle third contains your heart, and each outer third contains one lung. Thus the middle track of the DFL surrounds your heart, and also houses the various tubes within and around the heart and lungs. When the fascial webbing around these reaches the top of your lungs, it connects with the fascia of the Deep Front Arm Lines via 'the point in your pecs' (see Fig. 7.2 page 135).

This means that if you can access the Deep rather than the Superficial Front Arm Lines (thumbs on top!), and you have a well-functioning DFL, *your arms connect directly to the force that is being transmitted from your back to your front via your diaphragm and the webbing within and around your heart and lungs.* Myers calls the Deep Front Arm Lines 'the expression on the DFLs in the arms'[3]... It is a wonderful feeling when this connection adds to your ability (already enhanced by the Functional Lines and Lateral Lines as they too pass through the 'points in your pecs') to keep your arms still and push your hands forward!

The most forward track of the DFL finds its way right over the diaphragm to its attachment at the inside of the bottom of the sternum, and continues underneath this bone to your neck, where it lies in front of the posterior and middle tracks. By adding your diaphragm to the bolts in your pelvis (see exercise page 73) and your arm cuffs (see exercise page 145), you can now transmit force from your back to your front at all three levels of your torso.

1. Do you transmit force less well from the back to the front on one side of your body? This is likely to be the side that you rotate towards, with an increased temptation to pull on that rein. However, your pattern may not be rotational; it may be that this side of your torso just lags behind the other and is more 'soggy'.
2. In your chair and in the saddle, put 'both boards on' (see page 107) and think of the line they would take from the back of you to the front of you, through your diaphragm and insides.
3. On the more difficult side, think of getting the DFL to reach all the way over the dome of the diaphragm. Imagine stretching the last corner of a fitted bed-sheet over the corner of a (dome-shaped) mattress.
4. Put your first two fingers either just below your kneecap (if you are sitting), or just to that side of the pommel when riding, and pull on them, thinking of the bed-sheet as you do so. Can you feel that side advance and strengthen? This might also help you to keep the QL 'on the fence' on that side: push on the back of the saddle on that side to firm it up (see Fig. 3.15(b) page 72).

5. Previously, we have used rotation to adjust your torso (remember the ice-skater of Fig. 8.11 page 178). What you are doing now is *rotating against a resistance*, which creates a shear force. This is inherently stronger. Also, realise that, whichever side you shear ahead, your skewers keep pointing along the tangent of your circle (see Fig. 9.15).
6. In trot, the passage of energy over your diaphragm is constant rather than pulsing. The more aware you are of this, the more effectively you will match the forces of the horse's movement. Keep checking in on your weaker side.
7. If the horses you ride tend to break back to trot instead of maintaining canter, this central part of your DFL is involved. Think of each canter bound like a dolphin jump. You have to jump with it, and you do this via the energy that

Fig. 9.15 There is a significant difference between turning the torso so that one side advances, and shearing it so that one side advances. The latter puts one of the outer thirds of the torso ahead of the middle third, which in turn is ahead of the remaining third. Notice that if you shear to adjust your torso, the skewers of Fig 5.7 page 103 will always point straight on, rather than into or out of the circle. When you can reverse your natural pattern there will be huge improvements to your sitting and steering.

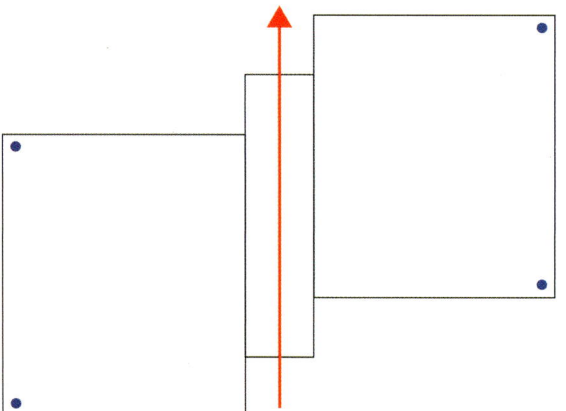
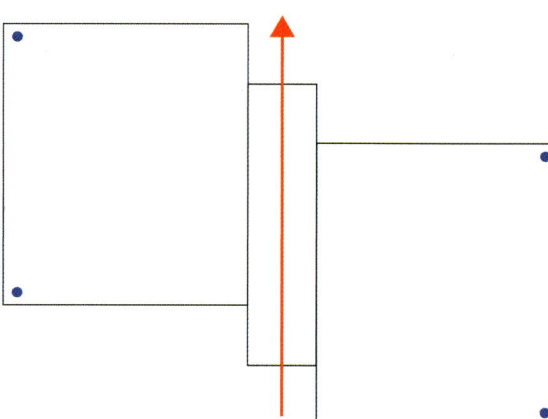

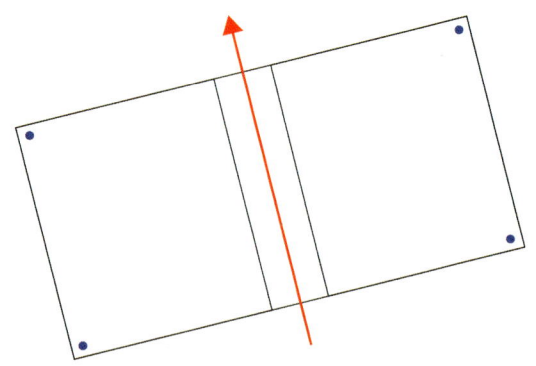
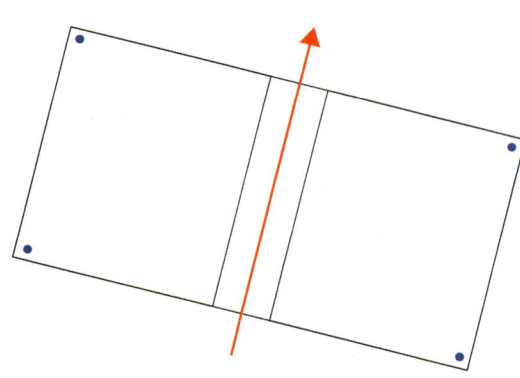

passes over your diaphragm along this track of the DFL. If the horse breaks back to trot, your energy got stuck at the back of your diaphragm. You may easily 'freeze' your energy here, especially if you have a sudden surprise.

This line of the DFL has connections to the muscles surrounding your hyoid bone – the bone that floats under your chin just above your voice box – and from there to your jaw and the temporal bone at the side of your skull. The hyoid bone is involved in swallowing, and lies at the root of the tongue, so the DFL has formed a line of pull connecting your tongue to your toes!

It is not only horses who do odd things with their tongues (they have very different hyoid bones from us); you too may find yourself holding your tongue in a strange way when you ride. The tongue is the most forward (and articulate) muscle of the DFL, and its contortions can be the source or the result of problems anywhere down the line. Many problems with riding and other sports requiring high concentration can be helped significantly by relaxing the tongue, and by extension of the jaw and the throat.

Remember, however, that you pushed your tongue against your molars to help you put more 'stuffing' into one side of your torso (see Fig. 5.16 page 114), engaging your diaphragm and your psoas on that side. This increased the pressure in that side of your body, and showed that relaxation is not always the answer!

Connectors and Spacers

The DFL as I have described it includes the fascia that surrounds your lungs, heart, and abdominal organs. This means that you can invent connections that go from anywhere to anywhere within your torso-box, allowing you to shore it up its weak places. This makes the DFL far more than a corset of muscles – it's functioning also depends on what is known as the turgor of the organs. Cells are like water balloons, and high turgor (or pressure) inside them helps your organs to hold up the inside of your torso-box. A wilted lettuce leaf has lost its turgor, and you are much better off without wilted organs inside your torso!

If any part of your torso-box becomes soggy, has its own little jiggle, or feels disconnected from the rest of the box, you can firm it up by connecting it to another part that is more stable. Think of these connections like girders within a modern building, which can have various angles and connect different levels. If two parts need to be connected together, create a 'connector' in your mind. If one part needs to be kept further away from another part, make a 'spacer'.

These allow force transmission along routes that were previously unavailable, and Fig. 9.16 overleaf shows the torso-box model that I use to illustrate some of my favourite connectors and spacers. Realise that if you cannot transmit force effectively from your back to your front, you have no chance of connecting your horse's quarters and loins to his forehand! It is sad but true that most riders' bodies are jiggled and deformed by the forces of the horse's movement, which means that they, in turn, are deviating, dissipating or deadening those forces instead of riding easily on them.

Fig. 9.16 Some possible connectors and spacers within the 'box' of the torso.

The Fan of Muscles

There is just one group of muscles that have been left out of our entire discussion so far, and that is because they resist classification within any of the Anatomy Trains lines. Their fibres run horizontally *across* the back of the pelvis, so whilst their fascia is definitely continuous with the DFL near the seat bones, this change in the direction of the muscular 'grain' breaks the usual Anatomy Trains rules. However, they are still best considered as a branch of the DFL, and along with other muscles they form a fan around the hip joint (see Fig. 9.17).

There are six of these small but important muscles on each side of the sacrum. You may have heard of the piriformis, the topmost one, as it is often associated with sciatic pain, and when mal-aligned it gives people grief. (The smart practitioner, however, would focus on the equivalent muscle on the *other side*, since the non-painful one has often 'stolen the bedclothes' from its poor complaining partner.)

The lowest muscle of this group (the quadratus femoris) goes from the outside of your seat bone to your greater trochanter (see Fig. 5.2 page 93), just under the lower portion of your gluteus maximus. The shape of the male pelvis means that

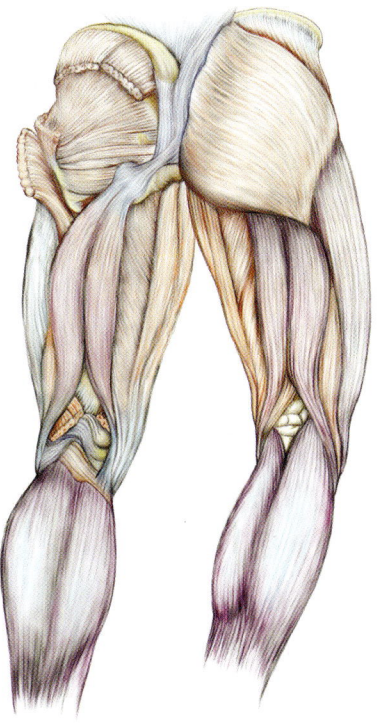

Fig. 9.17 The fan of muscles is visible under the cut-away buttock muscle in the left side. Some of them are also visible in the side view of Fig. 9.1 page 184.

it naturally becomes part of a man's sitting surface. For most women, sitting as described below is a learned skill which hugely improves our ability to organise the horse's back.

1. First on a chair, then on a saddle, pick up the flesh *at the side* of your backside, and pull it out to the side. Your seat bones become wider apart, and they clearly make contact with the surface you are sitting on. The diamond shape of your pelvic floor is also pulled wide.
2. Imagine drawing around the shape on each side that carries significant weight. Do you have two equal circles?
3. Now pick up the flesh at the side of your backside and tuck it in underneath you. Let your thighs make a more of a 'V' shape as you do this.
4. Are your seat bones more or less clear than before? Are they closer together or further apart? How does the shape of your sitting surface compare in the two positions? Imagine that you could draw around the area on each side that carries significant weight. How is it different? What has happened to the diamond of the pelvic floor?
5. Repeat this a few times, making your comparisons as rigorous as you can.
6. When your flesh is tucked in under, your seat bones are closer together and less clear. You are sitting on a bigger surface area to the *outside* of each seat bone. You may even be able to feel your lesser trochanters come down onto the chair.
7. Ideally, the quadratus femoris, the muscle that forms the bottom of the fan by connecting your seat bone and greater trochanter, lies across the horse's long back muscle on each side, as in Fig. 5.8 page 104.

8. Do not start working with this on your horse until you are ready to tackle sitting trot, and have your horse's Superficial Back Line organised well enough that sitting well is viable. When you take trot, sense your underneath: do your seat bones stay down on the saddle, or do they bounce up and down within your flesh? If the latter, your breeches may stay in place, but despite this you suffer from the 'spongy buttock syndrome'. (Sorry!)
9. Before you take trot, tuck your flesh in under on each side. (Do not tuck under from the back.) Think of pulling yourself *down* onto this surface, from your shoulders, elbows, and ribs. How many strides does it take for your seat bones and flesh to bounce out of this position?
10. Tuck the flesh in under, and begin again.
11. Be prepared to do this over and over, even if you have to stop each time after a few strides.

I have seen many people who thought they would never 'get it' gradually become able to maintain this way of sitting for progressively more time on a progressively more powerful trot. Your patience and diligence will be rewarded by much more efficiency; as you decrease the 'noise-to-signal ratio' in your sitting you gain a much clearer sense of your horse's back – and your horse will gain a much clearer sense of you!

The Deep Front Line in the Horse

The Danish researchers Vibeke Sødring Elrønd and Rikke Schultz have, at the time of writing, yet to formally report their findings on the DFL in the horse, but have discovered that it runs largely as it does in humans, with the proviso (as with the other lines) that the horse lives in hip flexion whilst we live predominantly in hip extension (though not when riding).

The DFL runs up the inner seam of the horse's hind leg, connecting the long flexor tendons via the adductors to his seat bones and pelvis (see Fig. 9.18). The connection to the psoas muscle is strong, such that less-than-sterling farriery can leave an angle on the hind hoof that leads to bucking, and affects canter transitions in particular. The horse's psoas becomes irritated when he has to step too far over the end of his overly long 'toenail'.

Skilled bodyworkers can work directly on the human psoas, easing their fingers through more superficial layers of the belly. On horses they can only reach it indirectly, through working on the inner thigh muscles. When the horse steps back with his head lower than his withers he stretches his psoas – and if he is reluctant to do so, it may indicate that he is compromised here.

Just as our psoas and iliacus muscles lie within our pelvis, so the horse's lie underneath his pelvis (which has a much flatter shape than ours). The iliopsoas complex is what enables him to flex the joint between his lumbar spine and his pelvis, lowering his croup and bringing his hind legs more under him. After he has learnt to draw his withers and ribcage up between his shoulder-blades, and to move

Fig. 9.18 The Deep Front Line in the horse connects his hind hooves to his skull, tongue, and hyoid bone. Within his torso we can think of the Deep Front Line having three tracks as we do in ourselves. It runs beneath his spine, and above his abdominal wall, with each of those tracks continuing through his neck and skull. The middle track includes his psoas complex and the organs of his abdominal cavity, his diaphragm, heart, lungs, trachea and oesophagus, making its way to the roof of his mouth. In the traditional idea of the 'ring of muscles', the scalene muscles, which join the first rib to the neck vertebrae would close the ring in his chest.

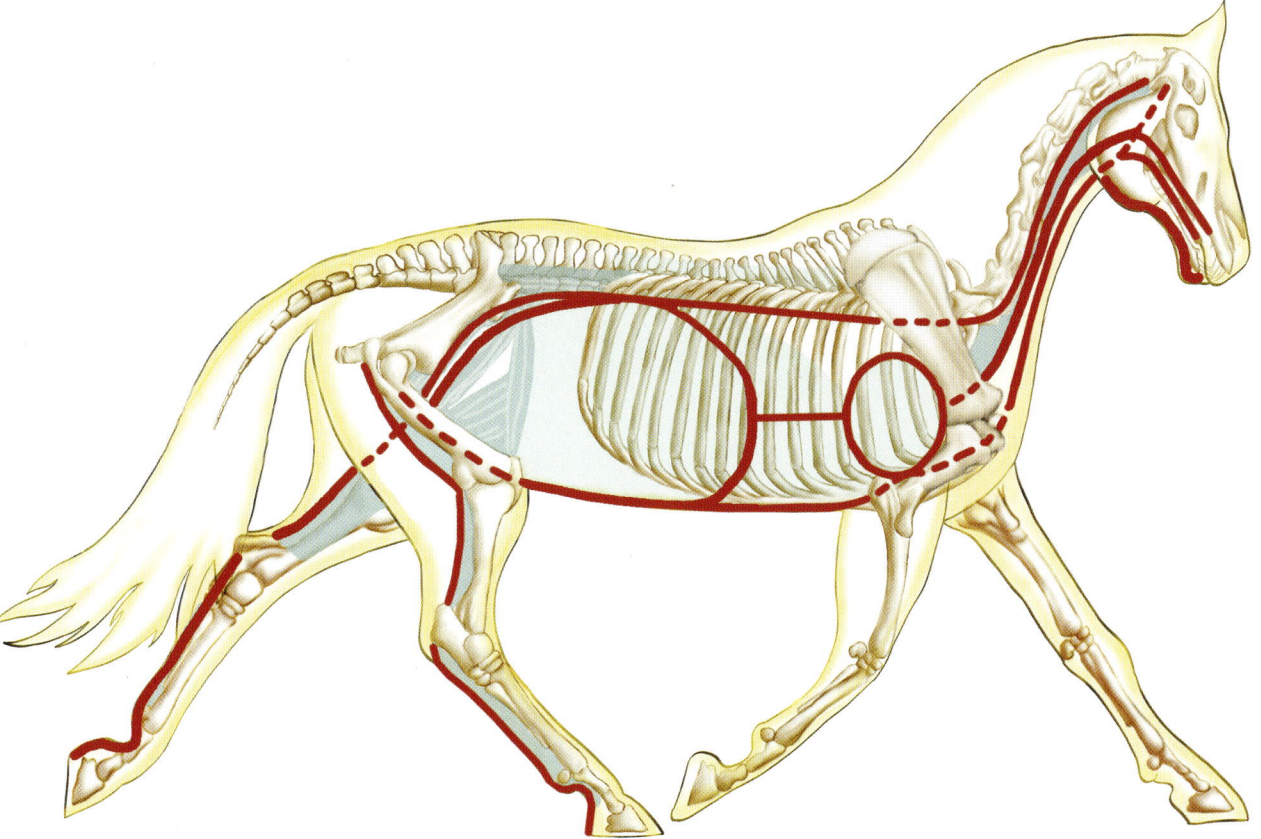

'like a gymnast on a balance beam' (see page 177), the DFL adds the ability to 'coil his loins'. This is the next tool in his 'toolbox' for collection.

Also like us, his psoas muscles connect to his diaphragm. Research veterinarian Rikke Schultz believes that restrictions in the diaphragm create many issues in both animals and people that are seriously under-diagnosed. There is more at stake than breathing well: if the various tubes that pass through it (for example the aorta, oesophagus and vagus nerve) are constricted, the well-being of the entire organism is affected. This may not be discovered in a pre-purchase vet check!

Our vertical posture means that our heart rests on our diaphragm, but the horse's rests on his sternum, slightly changing the configuration of the middle branch of the DFL. From his diaphragm the DFL passes through the membrane that separates his lungs, and the bag that surrounds his heart. It then continues via the tubes of his trachea and oesophagus into the base of the skull and the roof of the mouth.

The ventral (stomach) branch of the DFL passes along the belly wall through the very strong fascia that supports the guts and lungs. Some trainers doubt that the horse's abdominal muscles are strong enough to act as the 'bowstring' that helps to lift the bow of his back (see page 84); however, the DFL adds another layer of abdominal strength. It continues along the underside of his neck via muscles that attach to the hyoid bone, the tongue, and the bottom of his chin. The hyoid bone itself extends upwards from under his tongue to just below his ear, where our much smaller one has no such attachment. (Instead it floats just in front of our voice box.)

As in us, the horse's DFL has ultimately connected his tongue and hyoid bone to his hind 'toe'. Remarkably, if you put downward pressure on a bit placed in the mouth of a horse cadaver, the hind leg becomes restricted in its movement![4] For us as riders, this should be sobering. Perhaps less surprising (now that we have come this far!) is that, if your horse sticks his tongue out to one side, his problems could lie far deeper than his teeth or jaw (and attendant bitting issues), originating instead in restrictions further down the DFL that you would probably never suspect. Whilst some horses stick their tongue out as part of their evasive pattern and put it in when they work well, others function the opposite way around. Each case is a consequence of the way that good work engages the DFL.

Core to Core?

The big question is: can you, as a rider, use your core to influence your horse's core? Elite dressage riders undoubtedly do this, and I hope I have shown you, as this book has evolved, that it is a learnable skill. The key requirements are that you can (a) sit still, (b) match the forces of the horse's movement, and (most importantly) (c) project your attention via your own insides to your horse's insides, imagining muscles that lie deep within his torso and neck.

Connecting to Your Horse's Psoas Muscles

1. Your horse's psoas muscles connect into his diaphragm, which lies under the back triangle of your pelvic floor when you are riding. They pass diagonally to the rear and down, not actually touching his pelvis, and inserting into his thigh bones just below each hip socket.
2. Think of a connection from the underneath of your torso-box into those muscles, which would enable you to pull on his thigh bones, drawing his pelvis and hind legs under him.
3. Whilst the above is probably the most powerful example of this, you can also think of strings from your front tendons (see Fig. 3.20 page 79) passing diagonally back and down to pull on the horse's hind pasterns, or strings from your upper chest passing through you and becoming a Pessoa strap around his quarters. Alternatively, those strings could go diagonally past your sides, like the straps of a back-pack, and extend around his quarters. From your lower back you could connect some strings to your horse's loins, and pull them up and towards you.

4. Become inventive! The ideal muscle use, coupled with your ability to visualise a connection, makes these strings – or at least the line of pull they represent – really exist. They are not just figments of your imagination! The generic idea is to think of connecting one place in your torso to another place in your torso, to a place in your horse.
5. Imagine that your horse has a button at the root of his neck, about where the ring of a martingale would hang. Can you imagine two strings connecting through his body to just above and just behind your pubic bone, continuing through to the back of you and connecting to two points near the top of your pelvis? Do you pull the root of his neck towards you, or does he pull your lower back forward?

The 'Ring of Muscles' and Longus Colli

We usually think of the horse's neck being hung from his topline (the Superficial Back Line), but it is also supported from beneath the neck vertebrae by the third (dorsal or back) branch of the DFL, which passes from tail to skull right beneath his spine, via his ventral longitudinal spinal ligament. This mirrors the DFL in us. Again as in us, the longus colli muscles lie to each side of the ligament, beginning under the vertebrae of his ribcage, towards the front of your pelvic floor. The longus colli takes the DFL from here to his atlas, the first vertebra of his neck.

At the root of the horse's neck the scalene muscles lie just to the outside of longus colli, again, as they do in us. They join his first rib on each side to the middle neck vertebrae. Together, these muscles make a sling under the lower neck vertebrae, so when they contact, the root of the neck is lifted (see Fig. 9.18 page 211).

Dr Deb Bennet believes that they are vital to what she calls the 'neck telescoping gesture' though which the neck reaches and arches out of the withers.[5] She includes them in the fabled 'ring of muscles' which influence the horse's posture. As traditionally described, this ring includes muscles that are part of the Superficial Back and Front Lines, and also muscles that are part of the DFL. But by moving between different layers, this schema breaks the rules of Anatomy Trains, and I am proposing a review the ring, adjusting it so that it functions entirely within the deepest level of muscle and fascia.

Realise that on our way to this point we have discovered two other rings that are more superficial. The Superficial Back Line and Superficial Front Line form a ring from the hind hooves to the masseter muscles of the jaw (see Fig. 4.3 page 83), and the Functional Lines form diagonal rings between opposite stifles and elbows (see Fig. 6.5 page 130). The DFL also makes a ring of deep myofascia, which closes rather like the ring of the Superficial Front and Back Lines, in the jaw.

The traditional ring of muscles is considered to close in the horse's chest, by passing from the abdominal muscles along the sternum to the first rib and thus to the scalene muscles. From our understanding of the DFL we would substitute the deep fascia of the belly for the abdominal muscles and the fascia above the sternum for the sternum itself.

This traditional ring is a compelling idea that would really facilitate collection, but unfortunately the evidence from the dissections done by research veterinarians does not support it. However, it is still true that when the horse engages his scalenes and longus colli, his first rib is pulled up towards his neck vertebrae, and his neck is supported from below. His carriage becomes that of the archetypal dressage horse, with a beautiful reach and arch to his neck.

You will not have to hold his neck up with your hands, because you are using your thighs and pelvic floor to tell your horse to hold it up himself! In addition, his psoas muscles have lowered his croup and bought his hind legs under his body, which cantilevers his withers up. This means that the withers have been lifted from above, from below, and also via the muscle sling!

1. Imagine the longus colli muscles, and how they support those neck vertebrae. This thought might change how you organise your pelvic floor, the underside of your upper thighs, and your front.
2. Be prepared for a different feeling of 'there-ness' on the end of the rein – light but distinctly present. It is easy for many riders to get the horse too light as he backs away from the bit – this way of him seeking contact is a very different feeling, as his neck supports its own weight, but elongates into the contact.

The 'Treadmill'

A more general way of thinking about the horse's core is as follows. Think of a plane that lies horizontally within your horse's torso, at about the level of your knees, or slightly below them if you are a short rider. Make this into a horse-shaped treadmill, that goes all the way from the horse's seat bones and hindquarters to his chest and the root of his neck: keep your focus on the way that it rolls forward along its top – realising that it also turns back under itself as it returns to his hindquarters.

1. Does the treadmill go all the way from the back of him to the front of his chest, or does one end of it stop short? Does it roll forward without any glitches? Does it feel the same in both sides of him? If not, can you change this?
2. When you go from trot to walk, imagine it changing smoothly from a trot treadmill to a walk treadmill.
3. Use this principle for any transition, and see if this helps you to make it smoother, and to resist the temptation to make all the traditional mistakes of page 144.
4. Potentially, you could collect your horse by thinking of the front of the treadmill being raised a few degrees!

There is no end to how creative and effective you can be once you have discovered how your intention can 'think' an idea into life. But that is only possible once you are well stabilised and it is you who are organising the horse's strings, rather than him disorganising yours! By the time you and your horse are activating your Deep Front Lines, you will both have wonderful body awareness, remarkable focus and impressive tensegrity. The wonderful dance of 'equipoise' is a skill that is learned, as well as a gift that is given.

CHAPTER 9 NOTES

1. Myers, Thomas, *Anatomy Trains,* 3rd edn, Edinburgh, Churchill Livingstone (2014), p.185
2. Myers, Thomas, *Anatomy Trains,* 3rd edn, Edinburgh, Churchill Livingstone (2014)
3. Myers, Thomas, *Anatomy Trains,* 3rd edn, Edinburgh, Churchill Livingstone (2014), p.196
4. Sharon May Davis loves to demonstrate this in her dissection classes.
5. www.equinestudies.org/ring_revisited_2008/ring_of_muscles_2008_pdf.pdf

CONCLUSION – RIDING AS A LONG-TERM PROJECT

The suggestions in this book build progressively from the basic alignment that comes from balancing the Superficial Back and Front Lines. I hope you have had the magical experience of discovering how adjusting these lines in yourself progressively yields the skills to adjust them in your horse. The more skilled you become, the more far-reaching those improvements can be; yet the balance between these lines needs to be reviewed regularly. Most of us easily lose it.

We then embarked on equalising the Lateral Lines on both sides of both 'animals'. Human and equine asymmetries so often compound each other, and working with them is complicated by the rider's instinctive desire to turn by pulling on the inside rein. The horse turns most instinctively by falling in, but our human instincts – aided by the effects of the centrifugal force on the circle – get most partnerships into the biggest trouble. The answers become clearer as the Lateral Lines become more equal on each side, and the rider learns to steer the horse's withers instead of his nose.

Together, the lines along the front, back and sides constitute the 'outer sleeve' of rider and horse. The 'X's of the Functional Lines add more stabilisation to both the upper body and legs, and we discovered how their attachment to the opposite pelvis and thigh stabilises the Arm Lines. These also attach to the Lateral Lines, the Spiral Lines and the Deep Front Line, in ways that determine how the arms themselves operate. Pushing the hands forward also requires the ability to match the forces of the horse's movement, just as you would match the forces of a 'hippity-hop'. All of this confirms what most of us knew from experience: there is more to the control and placement of the arms than the arms themselves!

The Spiral Lines and Deep Front Line yield answers for more of the issues inherent in sitting well, and developing the skills to mould the horse's body and movement. The Spiral Lines build on the rider's ability to align her own and the horse's 'double yellow lines', and this new layer of skill unlocks the secrets of 'bend'. The rider has already discovered how her 'boards' can steer the horse's mid-line and cut down the wriggle room for his withers. Now her 'rebars' give her more ways to stabilise each corner of her torso-box, and to send forces through both of their bodies in the direction of her choosing. 'Inside leg to outside rein' has a new interpretation, and a new value.

Whilst elite riders make a profound 'core to core' connection with their horses, it is sad but true that, without ever meaning to, most riders mould their horses into a shape and movement pattern that is far from ideal. All of the myofascial lines can lead us to solutions, and through using the model presented in this book many of our problems really can be overcome. Through changing ourselves, we learn how to change our horses, balancing the myofascial lines of the rider-horse system.

Our biggest limitation is 'brain-space' or mental processing power. Consciousness is a blunt tool, but it is all that we have. It has been proved that there is no such thing as multi-tasking, so your only choice is to keep cycling through a limited number of triggers, checking in on them again and again. As some become automatic, this frees up brain-space, and you can take on more. Of the many ideas in this book, expect to be able to work with only a few at a time. The trick is to find the right few, at the right time!

Inevitably, the learning process requires time, dedication, determination, and focus – it mirrors the challenge of learning a martial art. However, very few people think of riding in this way – after all, it looks so easy! Older riders, who rode 'by the seat of their pants' as children and teenagers, return to riding after having children, etc., and discover that this approach no longer works for them. To feel secure and effective, they need skills (and of course, it is better to learn these skills when young). But few people encounter teaching techniques that are powerful enough to combat the force of habit, the pull of the myofascial lines, and the rider's instinctive mistrust of anything that feels weird.

A small proportion of the riders I meet have done a remarkably good job of teaching themselves to ride from my books, DVDs and webinars. That is a fantastic achievement, and this book – along with the resources that go with it – make it more possible. I do not underestimate the task; but I also know that any journey works best when you have a viable map of the territory.

Riding is at its most satisfying when it becomes an ongoing process of learning, and it can remain fascinating for life: there is no end to the way you can develop your body into a stable, adaptable tensegrity structure that transmits force. Your influence becomes increasingly profound and precise, and when you base your horse's training on the foundation of good biomechanics, your intentions become so much clearer to him. With less 'noise' and more meaningful signals, both of you develop your perceptions and body awareness. The prizes of this work may or may not lie in competition placings; but they undoubtedly lie in the many small but profound insights that build on themselves as your skill develops. They can enrich your life beyond measure, both on and off your horse.

May the force (from back to front) be with you!

GLOSSARY – THE 'LINES'

The following is a summary of the abbreviations for the various body lines used throughout the text. They are listed in order of the chapter in which they first appear.

CHAPTER 3
SBL	Superficial Back Line
SFL	Superficial Front Line

CHAPTER 5
LL(s)	Lateral Line(s)
DLL	Deep Lateral Line
SLL	Superficial Lateral Line

CHAPTER 6
FL(s)	Functional Line(s)
BFL	Back Functional Line
FFL	Front Functional Line
IFL	Ipsilateral Functional Line

CHAPTER 7
AL(s)	Arm Line(s)
DAL(s)	Deep Arm Line(s)
DFAL	Deep Front Arm Line
DBAL	Deep Back Arm Line
SAL(s)	Superficial Arm Line(s)
SFAL	Superficial Front Arm Line
SBAL	Superficial Back Arm Line
FLPL	Front Limb Protraction Line (Horse's 'Front Arm Line')
FLRL F	Front Limb Retraction Line (Horse's 'Back Arm Line')

CHAPTER 8
SPL(s)	Spiral Line(s)

CHAPTER 9
DFL	Deep Front Line
QL	Quadratus lumborum (muscle group)

INTERNET CONNECTIVITY

This book is only one of the ways in which I am sharing with you my knowledge of Rider Biomechanics. When you read, hear, see and do, your learning is infinitely enriched, and theoretical knowledge can more easily be transformed into practical skills. The website www.riderbiomechanics.co.uk has short videos of the exercises in this book, which are accessible via a password that you will be directed to find in these pages. www.dressagetraining.tv has longer webinars that show riders at various levels learning and demonstrating these skills. It also sponsors events for large audiences. My previous books and DVDs explain much of the same material, but with less reference to the fabric of the body. They can be sourced from www.mary-wanless.com, which also has information about my clinics in the UK and USA, and a list of accredited coaches worldwide. Immerse yourself in every dimension of this learning experience that you can; it will pay you handsomely.

INDEX

A

abdominal 'X' 167–168
Achilles tendons 26, 81
adductor muscles 125, 185, 186, 196
  in horses 131
adjustable stability systems 94
ALs *see* Arm Lines (ALs)
Anatomy Trains (Myers) 22, 25, 28, 47, 208, 213
Anatomy Trains (concept) 9, 10, 22, 42, 43, 47, 124, 162, 208, 213
ankles 51, 116, 189
areolar layer 18
Arm Lines (ALs)
  hands 141–144
  in horses 154–158
  passive resistance 144–145
  in riders 135–137
  shortening reins 145–148, 149–151
  wrists and elbows 138–141
asymmetry
  in horses 156–158, 175, 176
  in rider-horse interaction 23, 99, 101–104, 216
  and turning 170, 176

B

Back Functional Line (BFL) 122, 123, 127–128
ballistic training 87
bearing down 62–65, 71, 84
bend, in horses 96, 98–99
Bennet, Dr Deb 176, 213
BFL *see* Back Functional Line (BFL)
Blitz, Heather 25, 35, 41, 54, 88
Boards Exercise 107–111
body-mind 13, 36, 41–43
brain-space 216
breathing 62–65

C

'C' curves 100, 103, 111, 112–114, 168–170
calves
  cross-section 185
  imbalance in 49, 73–74
  position of 129
  and resistance 76, 77
  stabilisation 81, 89
  wobbly 189–190
carrying legs 97
chronic injuries, source 26
circles
  falling in on 115–116, 179, 216
  falling out of 114–115, 157
  fencing lunge position 178, 179
  ice-skating position 178, 179
  rebars 181–182
Clayton, Professor Emerita, Hilary 11, 98, 131, 156
concentric loading 30–31
connectors 207–208
conscious competence 40–41
contact scale, reins 148–154
core (definition) 184–188
core to core connection 212–214, 216
Coyle, Daniel 40
curling toes 81, 188–189

D

DALs *see* Deep Arm Lines (DALs)
DBAL *see* Deep Back Arm Line (DBAL)
deep and round position 85
Deep Arm Lines (DALs) 134, 137, 138, 154
Deep Back Arm Line (DBAL) 134, 135, 137, 161
Deep Front Arm Line (DFAL) 134, 135, 136, 137, 205
Deep Front Line (DFL)
  calves 188–189
  connectors and spacers 207–208
  core to core connection 212–214, 216
  diaphragm 202, 203–207
  fan of muscles 208–210
  in horses 210–212
  locals 200–201
  pelvic floor 195–198
  psoas muscle 186, 187, 198–200
  quadratus lumborum (QL) 193, 201–202
  in riders 184–188
  thighs 191–194
  toes 188–189

Deep Lateral Line (DLL) 95
deep practice 40–41
default balance patterns 54, 55
dehydration 28, 29, 32
dexterity 142, 156, 157
DFAL *see* Deep Front Arm Line (DFAL)
DFL *see* Deep Front Line (DFL)
diaphragm 202, 203–207
DLL *see* Deep Lateral Line (DLL)
dorsal line 82, 83
double yellow lines 119–120, 176, 216
Dove, Millie 11, 88
Dujardin, Charlotte 73, 145

E
ears, dominant 156
eccentric loading 30–31
elasticity 26
elbows 139, 140–141
equipoise 24–25, 35, 36–37, 214
Ericsson, K. Anders 39, 40, 41
expertise-induced amnesia 38, 43
eyes, dominant 156

F
falling in on a circle 115–116, 179, 216
falling out of a circle 114–115, 157
false confidence posture 54
fan of muscles (of DFL) 208–210
fascia
  definition 18–20
  elasticity 26, 32
  fascial connection 42–43
  fascial 'MELTing' 24
  fascial net 20, 26, 29–30
  fitness 24, 32
  in horses 24–25
  and hydration 28–31
  qualities 21–25
  in riders 24–25
  stretching 24
  strings 26
  tensegrity 25–28, 35
  types 33–37
fascia profundis 18
fascicles 29
feel 38–44
feet 81, 188–189
fencing lunge position 178, 179
FFL *see* Front Functional Line (FFL)
FLLs *see* Front Limb Lines (FLLs)
FLPL *see* Front Limb Protraction Line (FLPL)
FLRL *see* Front Limb Retraction Line (FLRL)
FLs *see* Functional Lines (FLs)
forelegs, role of 98, 155, 156
forward, down and out posture, in horses 85

front-back balance 47–52, 53, 57–62
  in horses 83
Front Functional Line (FFL) 122, 123, 124–125, 127–128
Front Limb Lines (FLLs) 154, 155, 156
Front Limb Protraction Line (FLPL) 154, 155
Front Limb Retraction Line (FLRL) 154, 155, 156–157
Fuller, Buckminster 25
Functional Lines (FLs)
  centring the cross of the 'X' 128
  in horses 19, 129–133, 216
  in legs 129, 216
  in riders 122–126, 216
  timing 127

G
girlish cast pelvis 94, 100, 112, 113, 164
Golgi tendon organs 42
grazing position 156–157

H
hamstrings 49, 52, 76–77, 186
hands 141–144, 216
head-carriage 165–167
Hester, Carl 87
hind legs, role of 97–98, 99, 156
hip joints, action of 54–55, 166, 202
holistic approach to training 31–32, 42
hollow-backed posture
  finding neutral spine 55, 66–68
  leaning back 56
  leaning forward 56
  and pelvic floor 196, 197
  without leaning forward or back 55
horses
  adductor muscles 131
  'Arm Lines' 154–158
  asymmetry in 156–158, 175, 176
  Back Arm Lines *see* Front Limb Retraction Line
  bend in 96, 98–99
  core 212, 214
  Deep Front Line 210–212
  fascia 24–25
  forelegs 98, 155, 156
  Front Arm Lines *see* Front Limb Protraction Line
  front-back balance 83
  Front Limb Protraction Line 154, 155
  Front Limb Retraction Line 154, 155, 156–157
  Functional Lines 19, 129–133, 216
  grazing position 156–157
  hind legs 97–98, 99, 156
  horse posture affecting rider 85–86

hyperextension in 85–86
jack-knife stance 95–96, 97, 98–99, 107, 114
Lateral Lines 94–99, 216
laterality 156
longus colli 213–214
neck telescoping gestures 213
psoas muscle 210, 211, 212–213
ribcages 97, 98, 99, 115
ring of muscles 213
shoulder-blades 154, 175, 176, 177
Spiral Lines 96, 97, 98, 174–183
Superficial Back Line 82–85, 96, 119–120
Superficial Front Line 83–85
topline 82, 213
as train 98, 176
ventral line 83, 84
withers 105, 107, 125, 176, 177, 216
young horses in training 97
hydration, and fascia 28–31
hydraulic amplification 192
hyperextension 85–86

I

ice-skating position 178, 179
IFL *see* Ipsilateral Functional Line (IFL)
iliopsoas complex 193, 200, 210
inner stability 14
inside leg to outside rein 101, 178, 181, 216
intermediate stability system 92, 99, 107–111, 119–120
interstitial nerve endings 42
Ipsilateral Functional Line (IFL) 129

J

jack-knife stance 95–96, 97, 98–99, 107, 114

K

Kegel exercise 197–198
kinaesthesia 14, 40, 41
kinaesthetic intelligence (KQ) 14, 40, 41
Kinesis Myofascial Integration (KMI) 23
kinetic chains 20, 25
KMI *see* Kinesis Myofascial Integration (KMI)
knees
  down more 74–75
  gripping with 191
  hyperextended 52
  neutral spine 70
  and toes 172, 189
  up and out in front 78–79
KQ *see* kinaesthetic intelligence (KQ)
Kuhnke, Sandra 157

L

Lateral Lines (LLs)
  double yellow lines 119–120, 176, 216
  in horses 94–99, 216
  in legs 116–120, 185
  narrowness 107–111
  in riders 92–94, 216
lateral shifts 164–165, 170–171, 183
laterality, in horses 156
leakage of Spiral Lines 162, 163
learning process 38–43
legs
  Functional Lines 129, 216
  imbalance in 73–74
  Lateral Lines 116–120, 185
  leg-pits 193–194, 196, 198, 200
  leg-yield 170, 182–183
  lower leg stability 124
  Spiral Lines 172–173
lengthening reins 150
lesser trochanter 104, 196
lines of pull *see also* myofascial lines
  definition 15
  importance of 14, 20, 21, 24–25, 26
  tension 34
  in training 28, 31, 43
LLs *see* Lateral Lines (LLs)
locals 200–201
locked long (definition) 30, 31
locked short (definition) 30, 31
long-chain training 32
longus colli, in horses 213–214

M

mantrap 86, 155, 194, 204
McLean, Dr Andrew 98, 157
mid-back, intractability of 202
muscle contractions 26, 29–31
muscle-isolation training 31–32
Myers, Thomas 9–10, 11, 20, 21–23, 24, 25, 28, 31, 34, 49, 83, 129, 193, 198, 206
myofascial force transmission 34–37
myofascial lines, role of 24, 36, 43, 46–47, 216–217 *see also* lines of pull
myofascial meridians map 9, 21, 22, 28

N

narrowness 104–111
neck telescoping gestures 213
nerve endings, and fascial connection 42–43
neuromyofascial web 21
neutral spine
  checking/maintaining 71–73
  definition 53–54
  finding 57–71

as a hollow-backed rider 55, 66–68
postural permutations 54–57
as a round-backed rider 56, 57, 68–71
strengthening 73
noise-to-signal ratios 14, 210
nuchal ligaments 82, 83

O

orange segments (of muscle) 20, 186, 192–193, 196
outer stability system 92, 112–114
outward stability 14
over the back posture, in horses 85

P

Pacini nerve endings 42
passive resistance 144–145
pelvic floor 195–198
pelvis
   associated problems 52, 141
   girlish cast 94, 100, 112, 113, 164
   and narrowness 105, 164
   and neutral spine 66, 68, 70, 71, 73
   spiral seat 178
peripheral vision 65, 166
piriformis 208
postural imbalance, observing 22, 23
psoas muscle
   in horses 210, 211, 212–213
   in riders 186, 187, 198–200
pulling on reins 141, 142, 151–154
pushing hands forward 141–144, 216
pushing legs 97

Q

QL *see* quadratus lumborum (QL)
quadratus femoris 208, 209
quadratus lumborum (QL) 193, 201–202
quads/quadriceps 32, 73, 77, 186

R

rami 195, 196
rebars 179–182, 183, 216
rebound elasticity 32
reinforcing bars *see* rebars
reins
   and contact 148–154
   horse putting a loop in 150
   lengthening 150
   pulling on 141, 142, 151–154
   as rods 154
   shortening 145–148, 149–151
relaxation 46–47
ribcages
   in horses 97, 98, 99, 115
   lateral shifts 164, 170–171
   and shoulder-blades 148
rider-horse interaction
   asymmetry in 23, 99, 101–104, 216
   'C' curves 112–114
   core to core connection 212–214, 216
   falling in on a circle 115–116, 179, 216
   falling out of a circle 114–115, 157
   going straight on 115–116
   narrowness 104–111
   side bends 99–100
   wriggle room 105, 107, 125, 176, 216
ring of muscles, in horses 213
rising trot mechanism 87–91
Rolf, Dr Ida 23
rotational preferences 165
round-backed posture
   finding neutral spine 56, 57, 68–71
   leaning back 57
   leaning forward 56
   and pelvic floor 196, 197
   without leaning forward or back 56
Ruffini nerve endings 42

S

'S' curves 57, 103, 163–164, 167, 170
sacroiliac 'X' 173
SALs *see* Superficial Arm Lines (SALs)
SBAL *see* Superficial Back Arm Line (SBAL)
SBL *see* Superficial Back Line (SBL)
scars, effects of 26, 28, 33
Schleip, Dr Robert 21, 42
Schultz, Rikke 9, 11, 19, 24, 154, 210, 211
seat bones
   heavier 168–169
   missing 102–104
   narrowness 104–107
   and pelvic floor 195–196
SFAL *see* Superficial Front Arm Line (SFAL)
SFL *see* Superficial Front Line (SFL)
shortening reins 145–148, 149–151
shoulder-blades 136, 137, 148, 163
   in horses 154, 175, 176, 177
shoulder-in 98, 132, 133, 170, 182–183
shoulders, rider's parallel to horse's 178
side bends 95, 96, 99–100, 164–165
SLL *see* Superficial Lateral Line (SLL)
Sødring Elrønd, Vibeke 24, 154, 210
soft sides 97, 101, 133
spacers 207–208
Spiral Lines (SPLs)
   abdominal 'X' 167–168

head-carriage 165–167
heavier seat bones 168–170
in horses 96, 97, 98, 174–183
lateral shifts 164–165
leakage 162, 163
in legs 172–173
in riders 160–173
'S' curves 57, 103, 163–164, 167, 170
sacroiliac 'X' 173
side bends 164–165
spiral seat 178
SPLs *see* Spiral Lines (SPLs)
stabilisation 46–47
steering, and withers 105, 107, 125, 176, 177, 216
stiff sides 97
straightness 98, 176
Structural Integration 23
stuffed posture (definition) 36
Superficial Arm Lines (SALs) 137, 138
Superficial Back Arm Line (SBAL) 135, 136, 154
Superficial Back Line (SBL)
  elongating 79–81
  in horses 82–85, 96, 119–120
  and neutral spine 53–62, 73
  in riders 47, 48–50, 185
  and rising trot 89–90
  shortening 74–78
Superficial Front Arm Line (SFAL) 134, 135, 136, 154, 205
Superficial Front Line (SFL)
  elongating 74–78
  in horses 83–85
  and intermediate stability system 119–120
  and neutral spine 53–62, 73
  in riders 47, 50–52, 185
  and rising trot 89–90
  shortening 79–81
Superficial Lateral Line (SLL) 95

T

Talent Code, The (Coyle) 41
Temple dancer fascia 33
tensegrity 25–28, 34, 35, 36, 47
tension integrity 25
tensional force 34–35

thighs
  balancing inner and outer 116, 191–194
  and bodyweight 75, 77
  imbalance in 73–74
  narrowness 104–107
  as three orange segments 186, 192–193, 196
thrust, and rising trot 90–91
tight is light 36, 71
toes 81, 188–189
topline, of horse 82, 213
training theories 31–33
treadmill (horse's core) 214
turning like a bus 107, 115, 176, 177
two-pack line 78

U

unconscious competence 40, 126
unfocused vision 65, 166
unstuffed posture (definition) 36

V

vagus brake 65
vagus nerve 65, 211
Valsalva Manoeuvre 63
Van der Wal, J. 42
ventral line, in horses 83, 84
vertical and neutral 54, 55, 61, 103
Viking fascia 33

W

wheelbarrow steering 107, 154
withers, as steering point 105, 107, 125, 176, 177, 216
wriggle room 105, 107, 125, 176, 216
wrists 138–141, 147

Y

young horses, imbalance in sides 9